Women in Engineering and Science

Series Editor

Jill S. Tietjen, Greenwood Village, CO, USA

The Springer Women in Engineering and Science series highlights women's accomplishments in these critical fields. The foundational volume in the series provides a broad overview of women's multi-faceted contributions to engineering over the last century. Each subsequent volume is dedicated to illuminating women's research and achievements in key, targeted areas of contemporary engineering and science endeavors.The goal for the series is to raise awareness of the pivotal work women are undertaking in areas of keen importance to our global community.

Shae Taylor • Mary Campbell
Editors

Women in Biopharma

Springer

Editors
Shae Taylor
Agilent Technologies
Mechanicsville, VA, USA

Mary Campbell
Agilent Technologies
Hillsborough, NJ, USA

ISSN 2509-6427 ISSN 2509-6435 (electronic)
Women in Engineering and Science
ISBN 978-3-031-80347-5 ISBN 978-3-031-80348-2 (eBook)
https://doi.org/10.1007/978-3-031-80348-2

This Springer imprint is published by the registered company Springer Nature Switzerland AG
The registered company address is: Gewerbestrasse 11, 6330 Cham, Switzerland

Introduction

The 2023 Nobel Prize in Medicine was awarded to Katalin Karikó, Ph.D., and Drew Weissman, Ph.D., for their work on mRNA vaccines. mRNA research was the basis for the Moderna and Pfizer vaccines launched during the SARS-CoV-2 (COVID-19) pandemic in 2020. In the growing field of biopharmaceuticals, women have made many contributions to the research, development, and delivery of these drugs.

Taking its beginnings from the chemical industry, the pharmaceutical industry was, for the better part of a century, based on medicines synthesized from chemicals. Early biopharmaceuticals had their start from cells and organisms. In 1976, Genentech became the first "biotechnology" company—a company created by a venture capitalist and a scientist. The company's research was based on the science of recombinant DNA (rDNA) [1].

The first FDA-approved biopharmaceutical was the human insulin Humulin®, developed by Genentech and launched by Eli Lilly in 1982 [2]. Since the development of Humulin, the creation of biopharmaceuticals has taken advantage of advances in protein, monoclonal antibody, and nucleic acid Research and Development (R&D) as well as manufacturing. The industry has flourished since then, with more than 400 biopharmaceuticals with current active licenses [3].

The roles for women in the field are wide-ranging and many. From bench scientists to pilot plant supervisors to women working with regulatory officials, there is a role for women in every sector of the industry. This book relates the narratives of several women working in the biopharmaceutical industry; they share their career stories, from education to their current roles. You will find that they require different educational backgrounds and often offer opportunities to expand your education for future roles. We hope this collection will encourage young women to pursue careers in the biopharmaceutical industry, as their contributions will be significant to the future of good health.

References

1. "Genentech: Our Founders," accessed March 31, 2024, https://www.gene.com/about-us/leadership/our-founders
2. Walsh G., & Walsh, E. (2022). Biopharmaceutical benchmarks 2022. *Nature Biotechnology, 40*(12), 1722–1760. https://doi.org/10.1038/s41587-022-01582-x
3. Quianzon, C. C., & Cheikh, I. (2012). History of insulin. *Journal of Community Hospital Internal Medicine Perspectives, 2*(2). https://doi.org/10.3402/jchimp.v2i2.18701

Contents

What Is Biopharma? Weaving the Web with Pandora's Box: Riding the Wave of a Paradigm Shift in Biopharmaceuticals

Sabrina Hawthorne

Contents

1 Foreshadowing Complexity in Biological Systems

Carl Woese wrote of "a new biology for a new century" in the so entitled article describing the paradigm shift from reductionism to a holistic approach [1]. Scientific approaches have gone from deducing a protein sequence by sizing degradation products with chromatographic methods in the 1960s to several Next Generation Sequencing (NGS) technologies in the 2000s and Third-Generation sequencing starting in the 2010s, providing astounding amounts of information readily at our fingertips [2]. Technological advances have enabled this integrated approach through systems biology with 'Omic-level data in several areas of analytical measurement, such as proteomic, metabolomic, lipidomic, and genomic. Collectively, many novel methods enable new levels of understanding cellular dynamics. In an endless search for the truth, science is full of twists, turns, and various nuances that perhaps have provided a driving force behind the wave of personalized medicine. Though the scientific community is empowered with new technologies to take this more holistic approach through systems biology, the emphasis is on the research side, with a focus on identifying new targets.

S. Hawthorne (✉)
Global Biopharma Segment – Agilent Technologies, Santa Clara, CA, USA
e-mail: Sabrina.Hawthorne@agilent.com

S. Taylor, M. Campbell (eds.), *Women in Biopharma*, Women in Engineering and Science, https://doi.org/10.1007/978-3-031-80348-2_1

Alongside advances in sequencing technologies came new methods for phylogenetic approaches to deduce an evolutionary tree of life to help understand the history and relationships between taxonomic branches of living organisms. Carl Woese's work, published in his self-described "Pandora's Box of Microbiology," changed the very infrastructure of our taxonomy in textbooks. Taking a molecular evolution approach, his work demonstrated the classification of living organisms fit into a three-domain system rather than a five-kingdom system, along with great appreciation for being outnumbered substantially by microorganisms (Fig. 1). Based on phylogenetic sequence analysis, four of the five kingdoms (Animals, Plants, Fungi, Protists) were grouped as a single domain, Eukarya, and what was formerly the fifth kingdom, Monera or bacteria, became two separate domains, Eubacteria and Archaea. Though in textbooks now, his work was initially ignored, and he was ridiculed by the scientific community. It took over a decade for his insights to even be acknowledged, and about 20 years for the Woesian tree of life to be included in every microbiology textbook, and longer to reach all general biology textbooks [3]. As he built the vertical phylogenetic trees of single-cell organisms, he realized that it just scratches the surface of the complexity of the biological universe, as parts of the tree become cloaked by horizontal gene transfer, a mechanism enabling response to environmental change. Evolutionary biology aims to understand the dynamics of an entire ecosystem and the selective pressures that lead to adaptation. By incorporating the sequence of genes into phylogenetic trees, adaptation can be measured and traced within and between taxonomic branches. Yet, as Woese uncovered, genetic code can also impact the very ability to adapt to the environment, which adds intricacy to uncovering how these ecosystems change over time.

Woese's testimony outlines the complexity of biology and the crossroads the community will face in the twenty-first century. His research characterized the role of horizontal gene transfer in the dynamics of cellular evolution, predominantly in the Eubacteria and Archaea domains. Essentially, there are several categories of gene types with various levels of criticality to the cells, with some genes being more essential for survival. As a mechanism to adapt to extreme environmental changes, these essential genes can be transferred by horizontal gene transfer under selective pressures (Fig. 2). The most common method of horizontal gene transfer is a

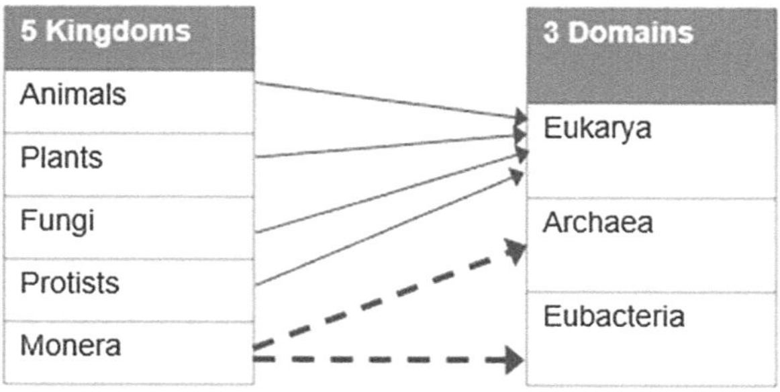

Fig. 1 The transformation of taxonomy from five kingdoms to three domains

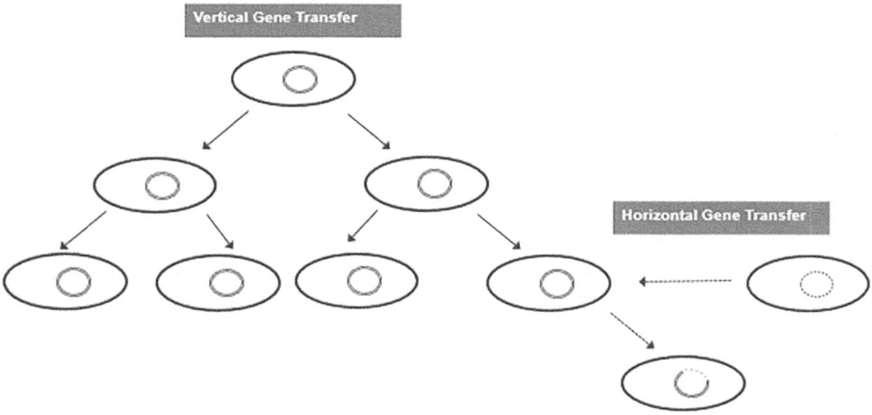

Fig. 2 The distinction between vertical and horizontal gene transfer

process called conjugation, which involves the transfer of nucleic acid on a plasmid or a small circular piece of DNA. The irony in this is that the harnessing of plasmids for recombinant DNA technology is what enabled the scaling up of mass production of therapeutic proteins such as recombinant insulin. This seemingly reductionist approach transformed our ability to treat disease with the introduction of biopharmaceuticals in the 1980s. Nearly a quarter of a century later, our toolbelt has expanded, and opportunities seem endless with new approaches in synthetic biology, protein engineering through artificial intelligence, and continual advances in gene-editing technologies. Further, with many advances in understanding the genome through methods such as single-molecule, long-read, or single-cell sequencing, we are still at the tip of the iceberg of this crossroad. As we continue to explore new levels of depth of information, insights are gained by balancing both reductionist and holistic approaches to treating disease.

With the continuing emergence of new technologies, we have yet to unravel the mysteries of how cells adapt to environmental changes. Transposable elements are another genomic structure connected with horizontal gene transfer and tend to contain repetitive sequences, making them difficult to map with short-read NGS methods. Interestingly, Barbara McClintock, the discoverer of transposable elements and Nobel Prize recipient, was also rejected by her scientific peers for proposing such a crazy notion of "jumping genes" [4]. Given the depth of research into how these transposable elements are involved in the ability to respond to stress, understanding these dynamics through long-read NGS could be another turning point in our grasp of genome complexity. The completion of the telomere-to-telomere project in 2022 marks a milestone in the quest to complete the picture of genomic architecture. Before chronicling several historical achievements that provide insights into the current landscape of the biopharmaceutical industry, a brief background on the cellular network and signal transduction is highlighted below.

2 The Cellular Network

Though the basic building blocks of life have been uncovered through reductionist means, the harnessing of these insights to elucidate and treat disease is still being pursued. The definition of life is centralized around the ability to respond to the environment, akin to water flowing in a stream and its ability to shift course should a stick or a rock be placed in its path. Any child lucky enough to enjoy the novelty of playing in a stream will quickly learn these rules of nature and may perhaps even imitate a beaver by attempting to block the entire water's path. The basic unit of life, the cell, operates in a similar manner, in that the parts work together to respond to the environment and maintain homeostasis through a rule structure of a vast network of interconnected signaling pathways encoded by the genome. Perhaps with the futuristic sick bay in mind, technology advances as does our ability to understand and visualize the biological response. Pushing the envelope, a generation of scientists influenced by Gene Rodenberry has enabled the next generation with tools at their fingertips to make this within reach.

Proteins are the biomolecules that do the work within a cell. Cells would not function without proteins. The genome of a cell contains the necessary information to produce proteins, and the human genome alone encodes for approximately 20,000–30,000 different proteins. Originally, it was postulated that each gene encodes a protein, termed central dogma; though, genes can code for several versions of RNA molecules or proteins. Protein-coding genes are transcribed by RNA polymerase to produce an mRNA that is translated by ribosomes, which fuse amino acids together by peptide bonds to produce proteins. Proteins are assembled by 20 different amino acids as encoded by the 64 codons within mRNAs. The exact order of the 20 amino acids determines the physical shape or structure of the protein and likewise determines function.

Proteins enable cells to interact with the external environment through the process of signal transduction. Through chemical interactions (covalent, ionic, hydrophobic, hydrophilic, or van der Waals), proteins are able to change their physical shape or conformation according to signals. This universal form of communication is also used to relay signals within an organism, as in the case of multicellular organisms, as signals could be from neighboring cell types. Inducing a physical conformation change of a protein by phosphorylation using either adenosine triphosphate (ATP) or guanosine triphosphate (GTP) is one of the ways in which signals are relayed. For simplicity, active and inactive are two basic types of conformations proteins can have, either enabling or preventing the cell to carry out a particular process. Small molecules can help to relay messages within a cell. Since they can easily diffuse through the cell, the signal transmits rapidly. These second messengers include molecules such as nitric oxide (NO), carbon monoxide (CO), cyclic adenosine monophosphate (cAMP), cyclic guanosine monophosphate (cGMP), calcium ions, diacylglycerol, and phosphatidylinositol. Collectively, proteins are the backbone of the cellular network, enabling its function. Signals from the environment or other cells are sent through this network, enabling a cell or organism to adapt to its surroundings. Illuminating the cellular network and its potential malfunctions is at the root of understanding and treating disease. With recent advances, we have reached the ability to measure perturbations in cellular dynamics using systems approaches and investigate tweaks to the systems to develop targeted therapies. The road to modern, molecular-driven, targeted approaches has been long and intertwined with regulatory oversight as we strive toward improving human health and the quality of life.

3 The Dawning of the Age of Biologics

Though there are various anecdotal stories throughout ancient history regarding medicine and our understanding of disease, as modern science has advanced, so has our approach toward the prognosis and treatment of disease. Historical accounts of at least two diseases, smallpox and diabetes, have ancient historical references, with many reports that have been written chronicling the details of early attempts at treatment. Smallpox is reported in mummified artifacts from Egypt dating to about 1400 BC [5]. The earliest written work describing symptoms similar to diabetes dates back to around the same time in ancient Egyptian papyrus records [6]. The synopsis of these two diseases below is an attempt to provide a framework for the establishment of modern-day biologics prior to the biopharmaceutical revolution.

Smallpox epidemics were frequent throughout the Middle Ages and impacted the development of Western civilization. By the eighteenth century, around 400,000 people died every year from smallpox in Europe. It was known as the "speckled monster," with a fatality rate as high as 60%, and survivors were often left with disfiguring scars and blindness [7]. Naturally, the desire to treat, cure, or prevent the disease was of significant interest. Using inoculations to protect against smallpox dates back to at least the Ottoman Empire in the 1600s and possibly earlier in China and India [8]. In the early 1700s, inoculation was introduced to England and involved taking pus from smallpox patients and infecting healthy individuals with small cuts. This form of protecting against disease came more from the practice of the people, rather than the scientific method, and had been passed down through the generations as folklore, going as far back as possibly the sixth century. What was originally described as superstition became so common that the medical community adopted the method. Unfortunately, many people died from the practice or contracted other diseases such as syphilis or leprosy [9].

A pivotal shift to intentional vaccination as a method to distribute immunity came from Edward Jenner's work, which used a scientific approach to testing the process of instilling immunity—laying the foundation for immunology. He had heard stories of milkmaids claiming to be resistant to smallpox after contracting cowpox, and to test this hypothesis, he used Variolae Vaccinae, or cowpox, to intentionally build immunity against smallpox—coining the term vaccination. He started his work in 1796, and by 1800, vaccination was adopted by most European countries—setting the stage for the biologics industry to come [7]. The nineteenth century saw additional vaccines for rabies, typhoid, cholera, and plague, prompting the establishment of vaccine farms. By the end of the nineteenth century, these had become prospering small businesses and were needed to keep up with demand. In the United States, most of the farms were operated by medical doctors who relied on simple cleaning measures and were likely inspected by local health officials [10]. For context, in the 1840s, Igor Semmelweis had worked to establish an association between unclean hands and childbed fever at the Vienna General Hospital. At the time, it was a fairly common practice for doctors to go from an autopsy to delivering a baby without washing their hands. Unfortunately, he was met with severe disbelief and criticism from his superiors and colleagues [11]. It was at least 20 years after his death when it was acknowledged that you really should clean up between the morgue and the delivery room. This came after Louis Pasteur's germ theory in 1861 became more mainstream and the aseptic surgical technique was established by Joseph Lister in the 1880s [11]. Quite fortunately, our understanding of the basic principles of sterility improved alongside our ventures into improving human health. The early part of the twentieth century brought additional vaccines for diphtheria, pertussis, tetanus, tuberculosis, and influenza, along with the additional need for the production of biologics [10].

Alongside the development of the biologics industry through vaccination and our understanding of basic principles of sterility, regulatory agencies were established to help maintain quality controls to ensure the safety of the public. The standard of drugs in the United States can be marked by The Vaccine Act of 1813, the establishment of the US Pharmacopeia in 1820, and the Drug Importation Act of 1848, which required the U.S. Customs Service to inspect drugs from overseas. After 13 children in St. Louis tragically died from a contaminated diphtheria antitoxin, Congress passed the Biologics Control Act in 1902. The passage of the Federal Food and Drugs Act in 1906, the 1938 Food, Drug, and Cosmetics Act, and the drug provisions of the 1938 Act led to what is now known as the Food and Drug Administration [12]. Though the regulation was initially focused on biologics for vaccine production, developments in the treatment of diabetes highlight another therapeutic utility for the up-and-coming biologics industry.

The chase for insulin in the pursuit of treating diabetes features increasing sophistication with novel methods in biologics on the path toward building the need and demand for the biopharmaceutical industry. Diabetes is first referenced in Egyptian medical texts from 1552 BC with a four-day treatment of a concoction including bones and other ingredients. The early reports of sweet urine, dating to at least AD 400 in India, were determined to be a result of the presence of sugar by Matthew Dobson in 1776. The time leading up to the discovery of insulin was also peppered with various medicinal treatments, including opium or dietary regimes [13]. There is a long history of diabetes research that is intertwined with the history of pancreatic anatomy and the characterization of enzymes in various extracts, starting with Paul Langerhans's description of what is now known as the islets of Langerhans in 1869. This was followed by Oscar Minkowski, Joseph von Mering, and Emmanuel Hedon's work on inducing a diabetic state in dogs, along with experimenting with pancreatic extracts for recovery. The push to isolate the factor in the pancreas responsible for lowering sugar in urine, using different extraction protocols, was pursued by numerous investigators, including Ludwig Zuelzer, David Scott, Israel Kleiner, Samuel Meltzer, and Nicolae Paulesco. The breakthrough came in the lab of Dr. John Macleod at the University of Toronto, with work done from 1921 to 1922 by Frederick Banting, Charles Best, and James Collip that resulted in a reproducible alcohol extraction method that removed toxic impurities and was effective in treating diabetes in rabbits and dogs [14].

Interestingly, for ambiguous reasons, of which there is still debate, only Banting and Macleod were awarded the Nobel Prize for the discovery of insulin, which inflamed an already antagonistic relationship, highlighting a pattern of behind-the-scenes sagas. To Banting's credit, he publicly acknowledged Best's contribution [15]. Their preparation of insulin then became the first therapy for diabetes in humans at the Toronto General Hospital [14]. This was groundbreaking and one of the first times that medical research was able to translate into a lifesaving therapy. The events that followed set the stage for the modern biopharmaceutical industry, including production scale-up and academic partnerships with industry. Further, the natural progression of biologics from vaccines to therapeutics has been demonstrated. Production facilities needed to be established to keep up with demand, much like the small businesses that popped up to meet the production demands of vaccines. Though, with the transition to a single molecule, more elaborate purification strategies were necessary for insulin.

After the success of the first insulin treatment in 1922, the next step was the scaling up of this purification method to meet the demands of patients with diabetes. The University of Toronto had experience with vaccine production and had established the noncommercial entity, Connaught Antitoxin Laboratories, in 1914, making it a logical partner to scale up insulin production. The public health entity operated in the basement of the University of Toronto Medical building and had been producing diphtheria antitoxin for vaccinations. After many attempts to scale up insulin production, the group sought outside assistance from Eli Lilly to overcome challenges with potency and stability. In 1922, Eli Lilly agreed to help develop the production methods with the University of Toronto in exchange for rights to manufacturing and distribution. By early 1923, they were able to produce enough insulin to meet global needs. The Connaught Laboratories was able to overcome production challenges by updating their equipment and facilities. There is a fair amount of speculation in the literature around why the team of scientists at the University of Toronto did not try to profit off the discovery, though it is clear that Banting and Macleod did not want to hinder the availability of insulin to those desperately in need. The team transferred their patent rights to the University of Toronto as a means to secure the technology from patents of potential competitors and in an attempt to prevent a commercial organization from creating a profitable monopoly. The University of Toronto empowered global production of insulin by granting patent rights to the Medical Research Council of Great Britain and the nonprofit Nordisk Insulin Laboratory, an affiliate of the University of Copenhagen [14]. Nordisk later merged with Novo in 1989 to form Novo Nordisk, a manufacturer of insulin as of this printing. Part of the requirement for receiving a license included a stipulation that the site meet criteria ensuring the production of insulin at the highest quality control standards, or good manufacturing practice, as determined by the insulin committee from the University of Toronto. Further, the University of Toronto controlled the right to batch testing from licensees in a test laboratory to maintain control of quality, which led to defining insulin standards. This continued until the 1941 Insulin Amendment, which required FDA certification of insulin production, with an exception made to the University of Toronto [16]. Ramping up the manufacturing of insulin, a biologic, required cooperation between academic institutes, nonprofit entities, and a commercial organization to meet the demands of the public to treat diabetes.

4 The Emergence of the Biopharmaceutical Industry

Over the next several decades, improvements to insulin preparation from pig and cow pancreas were made to obtain purer and more potent forms of insulin. There were still challenges with immune responses from patients due to the animal origins [17]. The sequence of bovine insulin was determined in 1955 by Frederick Sanger, marking the first protein sequence of amino acids [13]. Recognizing that insulin was a single protein meant that new measures could be taken to purify or even synthesize the molecule. By the 1970s, chromatography methods were developed that improved the purity of bovine or porcine insulin [17]. However, even with improved methods, the limited supply of animal pancreas was the bottleneck in the manufacturing of insulin until recombinant technologies opened the door to mass production of recombinant insulin, forming the biopharmaceutical industry.

The ability to measure insulin levels in the blood also underwent dramatic advances during this time with the development of immunoassays. While researching nondiabetic and diabetic metabolism of insulin, Solomon Berson and Rosalyn Yalow demonstrated the presence of anti-insulin antibodies [18]. The presence of these antibodies slowed the rate of insulin loss from the blood compared to subjects without insulin. This was followed by the development of a radioimmunoassay for monitoring insulin levels in the blood in 1961, enabling a highly specific and sensitive assay for insulin detection [19]. Prior to the immunoassay, insulin levels were based on the levels of glucose in the blood of depancreatized animals following injection of a sample or through in-vitro assays measuring glucose uptake by rat tissues, neither of which was particularly sensitive [18]. The breakthrough development of the radioimmunoassay for the detection of insulin paved the way for immunoassays for thousands of other analytes over the next half-century. At the time, antibodies were purified from human or animal sources, which led to demand for a steady supply of antibodies at an increased scale. By 1975, the ability to fuse a mouse myeloma cell line with a B-cell from an immunized mouse, enabling the indefinite growth in culture of a hybridoma cell line capable of producing large quantities of a monoclonal antibody, was described by Kohler and Milstein [19]. This milestone enabled a reliable source of antibodies against a specific antigen to be produced at scale for applications including immunodetection, immunopurification, diagnostic assays, research tools, cell marker identification, virus testing, and therapeutic applications [20].

The biopharmaceutical revolution came with the advancement of recombinant DNA. The ability to clone genes into plasmids, transfect them, and ferment microorganisms provided a manufacturing tool along with a biosynthesis process that produced superior purity and consistent quality. The FDA approval of recombinant insulin in 1982 was the defining moment for the biopharmaceutical industry. Rather than going through laborious extraction and purification methods of insulin, this harnessing of cells as manufacturing powerhouses brought a level of scalability to biologics that had not existed before, as depicted in Fig. 3. Though cells had been used previously, termed biologics, in vaccine production and originally for the purification of insulin, it was this recombinant method that enabled mass production or an increase in the scale of insulin production. Continual refinement and improvements to this method, using reductionist approaches, have ensured that biopharmaceutical development occurs at the highest quality standards as monitored continually through regulation. With the ability to produce recombinant proteins, monoclonal antibodies as therapeutics were not long to follow. With the biopharmaceutical revolution underway, the genomics era was soon at the doorstep.

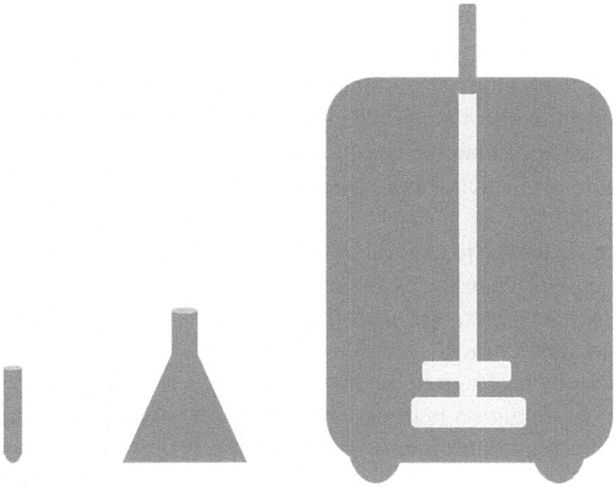

Fig. 3 Increasing scale of manufacturing enabled through recombinant technology

5 Monogenic Diseases into Superfluity of Oncology

Genetic disorders associated with a single gene are well characterized given their relatively simple inheritance patterns; yet intricacies arise as we examine nuances at the individual level, as strived for with personalized medicine. Beadle and Tatum's one gene, one enzyme hypothesis, proposed in 1941, followed by Vernon Ingram's work connecting altered hemoglobin and sickle cell anemia in 1956, served as the gateway to understanding mutations and their role in disease states. By 1966, the genetic code was cracked by Marshall Nirenberg, with a four-letter alphabet translating to 20 different amino acids. This was followed by two additional milestones in 1983: the first genetic mapping of a disease, Huntington's disease, and the invention of polymerase chain reaction (PCR), which paved the way for a new era in the approach toward both the treatment and diagnosis of disease. The quest for a complete map of human genes was seen as an atlas to treat disease. Further, detection and monitoring were becoming easier with new molecular tools.

The initiation of the Human Genome Project in 1990 was made possible by advances in sequencing technologies. The first working draft was completed by the spring of 2000, and a complete sequence was generated by 2003. With the human genome, efforts to map precise locations of genes became far simpler than the prior labor-intensive mapping methods. Further, the regions of variation between individuals, or single nucleotide polymorphisms, were more easily mapped [21]. This newfound understanding opened doors to innovative approaches toward the treatment and diagnosis of disease in a modern era of targeted treatment of disease based on the correlation between genes and the encoded protein function's role in the onset of the disease. The human genome provided a map to aid in targeted therapy of genetic diseases, enabling significant progress in treatment. How genomic information and targeted therapy have aided in the development of treatment for two diseases on opposite ends of genetic complexity is highlighted below. A synopsis is presented first for a well-characterized monogenic disease, cystic fibrosis (CF), and second for cancer, of which there are many types with complexities beyond the scope of this chapter.

Cystic fibrosis is a classic example of a simple autosomal recessive genetic disease caused by mutations in the cystic fibrosis transmembrane conductance regulator (CFTR) gene; however, the underlying mechanisms of disease onset are a complicated web with clinical variability and varied penetrance of CFTR variants. Since diagnosis is the first step in the treatment of disease, there is an inevitable pattern of advances in diagnosis being intertwined with advances in the mechanism of disease, though both arms are racing at staggering paces whilst trying to maintain parallelism. The discovery of the CFTR gene in 1989 enabled the development of a two-tiered diagnostic by Ronald Gregg in 1990, in which the immunoreactive trypsinogen detection takes place during newborn screening using a dried blood spot as the first tier, and a PCR-based molecular diagnostic as the second tier [22]. As of 2010, newborn screening is performed in all 50 states with a dried blood spot for a variety of diseases, including CF. The tiered approach is still used today, with the second tier often being molecular analysis for CFTR variants, for which a variety of tools are available. The CFTR gene is 189 kb with 27 exons and 26 introns, encoding a 1480-amino acid channel protein with two transmembrane domains, each connected to a nucleotide-binding domain.

The human genome, coupled with a functional understanding of the encoded proteins, set the stage for the development of targeted therapies. The CFTR gene encodes a channel that regulates the passage of chloride, bicarbonate, and sodium ions, which can lead to viscous mucus buildup and impairment. Understanding the role of the protein and the mechanism of disease has enabled therapeutics to be designed to target the causal mechanism of the disease. Highlighting the targeted therapeutic approach, small molecules were developed to modulate the activity of CFTR protein to treat cystic fibrosis, with clinical trials starting in 2006. Ivacaftor was the first drug to address a mechanism of action for cystic fibrosis by correcting the function of the CFTR protein and was approved by the FDA in 2012. It exemplified the use of a targeted therapy through the development of a drug that modulates the activity of a protein to treat disease. In the case of Ivacaftor, it was developed to impact a specific variant, 1652 G > A (nucleic acid change) or G551D (amino acid change); though studies have shown efficacy, safety, and effective treatment in additional mutations. The targeted approach to enhance or restore CFTR protein function through modulation also enables treatment of specific mutation classes. Additional research led to the development of Lumacaftor, Tezacaftor, and Elexacaftor in four different modulation therapies in clinical practice with indications specific to mutation type, as shown in Table 1. The most common mutation, the F508del, of which as many as 90% of CF patients have at least one copy of the variant, has pleiotropic effects depending on other cis or trans CFTR mutations, necessitating combination therapies with different mechanisms of action to rescue the phenotype.

Table 1 The number of approved CF mutations for targeted therapies in the United States (Roda et al., 2022)

Substance	# of CF mutations approved	Approved for F508del
Ivacaftor	38	No
Lumacaftor/ Ivacftor	1	Homozygotes
Tezacaftor/ Ivacaftor	26	Homo or hetero with additional mutations
Elexacaftor/ Tezacaftor/ Ivacaftor	1	At least one F508del

Alongside the development of targeted therapies, we have improved capacity to generate immense amounts of sequence data from patients. Though we have come a long way in terms of acquiring genetic data through next-generation sequencing methods, understanding the clinical utility of each variant and what the actionable options are is quite another journey. The use of next-generation sequencing has facilitated extensive profiling and collection of data from patients with cystic fibrosis, demonstrating that there are over 2000 mutations in the CFTR gene falling into six classes, of which around 360 are disease-causing [23–25]. As of 2023, a little over 30 years after the implementation of the two-tiered diagnostic, the FDA lists 12 different nucleic acid-based tests from eight companies, which can be used for in vitro diagnostics of cystic fibrosis. Further, the American College of Obstetricians and Gynecologists recommends carrier screening for CF for pregnant women or those considering pregnancy. As a direct-to-consumer test, there are countless clinics with their own carrier screening assays, in addition to offerings from commercial service providers. There are several types of CFTR genetic testing, including variant panels, traditional sequencing, next-generation sequencing, or deletion/duplication analysis, some of which have variable detection or sensitivities depending on the ethnic background and panel composition. The intricacy of the number of test types demonstrates the challenge posed to genetic counselors, which requires a wide skillset working alongside providers to interpret results and implications within the patient's medical and psychosocial framework. Additional complexities arise in that the panel used in carrier screening may not be the same as what is used in newborn screening [23]. Despite the challenges associated with interpreting and guiding through a complex decision process, it is essential for ensuring optimal treatment of CF with targeted molecular treatments. Further, the advances in detection have allowed for most to be diagnosed with CF prior to presenting symptoms, enabling preventative care. The case of cystic fibrosis presents a model in which the function of the encoded protein is well enough understood that monitoring a single gene, albeit a rather large gene, can facilitate a tailored therapeutic response. What about more complex diseases?

Though commonly perceived as a modern disease, cancer has roots that date back throughout ancient history. The earliest records of tumors are from 3000 to 1500 BC in ancient Egypt and include references to different types of cancers such as skin, uterine, and stomach. Prior to Marie and Pierre Curie's experimental approaches with radiotherapy in the late 1800s, the predominant approach to the treatment of cancer was through surgical or dietary measures. The first chemotherapies or drugs targeting tumors were developed between 1943 and 1948, with the first clinical studies using nitrogen mustard, better known as a chemical warfare weapon, in certain cancers such as lymphoid malignancies. The decades that followed used new therapeutic approaches such as antimetabolites, antimitotics, cytotoxic antibiotics, polyamine inhibitors, and iron-modulating drugs. By the 1960s, new combination chemotherapy regimens were used to improve therapeutic efficacy over a single drug [26].

The shift from untargeted to targeted therapeutics to treat cancer started with advances in cytogenetics. Though there had been speculation around the genetic origin of cancer, the tools were not available to establish the correlation until advances in cytogenetics in the 1950s. First, the confirmation of 46 chromosomes in normal human cells was established, followed in 1960 by karyotypes of patients with chronic myeloid leukemia (CML), which provided the first evidence of the genetic origin of malignancy. The chromosomal abnormality, termed the Philadelphia chromosome, was shown in 1973 to be the result of a translocation between chromosome 9 and 22, as the long arm of chromosome 22 (22q−) was shortened and the long arm of chromosome 9 was longer (9q+). By the mid-1980s, it was determined that the fusion gene BCR-ABL resulted from the translocation, which could be monitored through RT-PCR [27].

Advances in DNA sequencing, immunology, molecular, and cellular biology that followed enabled targeted approaches with a new class of drugs within biologics: therapeutic antibodies, which by definition and function are targeted [26]. The first therapeutic monoclonal antibody was approved by the FDA in 1986, which targeted CD3 as an immunosuppressant in transplants. Challenges with immunogenicity and potency led to the development of techniques such as chimeric or humanized monoclonal antibodies, leading to the 1994 FDA approval of the first chimeric antibody, abciximab, targeting GPIIb/IIa on platelets to treat clotting [28]. The first FDA-approved monoclonal antibody for oncology was Rituximab in 1997, targeting CD20. This new wave of biologic-based therapy was accompanied by regulations from the FDA, including the 1997 "Points to Consider in the Manufacture and Testing of Monoclonal Antibody Products for Human Use," which remains an important reference [26].

Shortly after a correlation between a gene product and its function was established, targeted approaches were used to develop therapies for cancer. After the HER2 gene product was determined to be a growth factor receptor and it was linked to breast cancer, a receptor antagonist drug development approach was taken to modulate the activity of the receptor. Herceptin® or Trastuzumab was the first targeted monoclonal antibody targeting an oncoprotein tested in clinical trials in 1992 against the HER2/neu receptor for breast cancer and approved in 1998. Of lesser recognition was the simultaneous FDA approval of the first companion diagnostic, HerceptTest, to detect HER2 expression in tumors. An immunohistochemistry (IHC) assay had been used throughout the clinical development process to detect HER2 overexpression in tumor samples for patient selection. Shortly after the start of Phase 3 clinical trials, the development of HercepTest, a robust standardized assay for routine diagnostic labs, was developed for detection of HER2 in tumor samples [29]. Importantly, Herceptin may not have been approved had it not been for patient selection. This targeted approach of a monoclonal antibody against an oncoprotein changed the treatment for the 20% of women with HER2-positive breast cancer. Not only could we develop targeted therapies, but we could identify the correct patients who would benefit from a particular therapeutic, signifying the beginning of precision medicine [29].

The continued pursuit of targeted therapy to treat cancer has cascaded with many novel options and opportunities for treatment. As Herceptin was approved, another wave of targeted therapies entered clinical trials in 1998, including the first of the tyrosine kinase inhibitors, a small molecule, Imatinib, also known as Gleevac®, which received FDA approval in 2001. This was followed by many molecular targeted therapeutic agents. More recently, genome sequencing is used to compare normal and tumor states, enabling the identification of new targets and biomarkers. Currently, the predominant types of targeted molecular therapy use small molecules or monoclonal antibodies to modulate growth factors, signaling molecules, cell-cycle proteins, angiogenesis, and others [30]. These fall into two major categories: either intrinsic to the tumor by targeting it directly or extrinsic to the tumor by targeting the surrounding microenvironment. The extrinsic factors can also be at an organismal level, such as the gut microbiome, of which emerging evidence has suggested a role in cancer development. Expanding the toolbox of therapeutic modalities, the biopharmaceutical industry is exploring a variety of novel drug types, including but not limited to gene therapies, cell therapies, and oncolytic viruses. In the case of hematologic cancers, the advancement of immunotherapy using chimeric antigen receptor-T (CAR-T) therapies exemplifies the utility of harnessing powerhouses extrinsic to the malignancy [31]. As of 2023, there are six approved CAR-T therapies, as seen in Table 2. Currently, a plethora of different cellular therapies are being pursued using both immune and stem cells to treat hematologic cancers and solid tumors, along with other disease types.

One of the hurdles in oncology associated with targeted therapy is the advent of drug resistance, which can be attributed to the heterogeneity of the cells that make up a tumor. An emerging class of cancer targets comes from a more holistic view of interpreting oncogenic cell states as a combination of intrinsic and extrinsic factors,

Table 2 FDA-approved CAR-T therapies

CAR-T	Target	FDA approval
Kymriah® (tisagenlecleucel)	CD19	2017
Yescarta® (axicabtagene ciloleucel)	CD19	2017
Tecartus® (brexucabtagene autoleucal)	CD19	2020
Breyanzi® (lisocabtagene maraleucel)	CD19	2021
Abecma® (idecabtagene vicleucel)	BCMA	2021
Carvykti® (ciltacabtagene autoleucel)	BCMA	2022

with the therapeutic opportunity to, in essence, tweak the system. Harnessing of the immune system to combat cancer with extrinsically targeted therapies such as checkpoint inhibitors and adoptive cell therapies has been pivotal for the treatment of a variety of cancers. The FDA approval of checkpoint inhibitor therapies, mAbs Nivolumab and Pembrolizumab in 2014, was only the beginning of our journey into modulating the tumor microenvironment or extrinsic factors. Along with qualifying new targets that can influence these heterogeneous states, markers will need to be defined to enable the stratification of patients [32].

6 Closing Remarks

Since the sequencing of the human genome, the prediction of the number of druggable targets has been in flux. Given the reductionist nature of science, as described by Woese, this number has been finite and a subset of the total number of genes coding for proteins. To granulate further, this can be pinpointed to specific amino acid residues in binding pockets of endogenous proteins with activities that can be modulated through interaction with either small molecules, antibodies, proteins, or peptide-based therapeutics. Beyond just the theoretical aspect of this comes the practicality of being able to deliver the therapeutic compound to the appropriate cells within the body without any unintended interactions or binding with negative consequences. However, the landmark FDA approval in December of 2023 of the first clustered regularly interspaced short palindromic repeats (CRISPR) therapy, exagamglogene autotemcel (Casgevy®), targeting a transcription factor, changes the paradigm of druggability. Transcription factors were previously deemed undruggable, though with new tools such as gene editing, anything is targetable. Coupled with the concept of viewing tumors as dynamic, with multiple cell states as influenced by intrinsic and extrinsic factors, this paints the picture for a multifactorial approach to treatment over a single target. The ability to complement intrinsically targeted therapies in combination with extrinsic agents provides hope for a diverse collection of regimens for precision medicine-guided treatments for patients.

The reductionist approach has transformed our ability to treat disease, and we are just beginning our ascent into approaching the system as a whole. The biologics industry transformed significantly through the nineteenth and twentieth centuries, addressing the scale of manufacturing through the rise of vaccine farms, production facilities, and advances in recombinant technology, leading to the biopharmaceutical industry. With advances in genomics and immunology alongside developments in understanding mechanisms of disease, the modernization of the ability to deliver targeted therapies continues, as summarized in Table 3. Approaches using

Table 3 Timeline of major milestones highlighted in chapter

Date	Milestone
BC	References to smallpox, diabetes, and cancer
1800	Vaccination adopted across Europe
1813	The Vaccine Act implemented in the United States—Beginning of Regulation of Biologics
1861	Louis Pasteurs' Germ Theory
1922	First successful treatment with insulin
1941	One Gene, One Enzyme, proposed by Beadle and Tatum
1955	Sequence of bovine insulin by Frederick Sanger
1956	Altered hemoglobin linked to sickle cell anemia
1961	Radioimmunoassay developed to monitor insulin levels in the blood
1973	Discovery of Philadelphia Chromosome linking genetic anomaly to cancer
1975	Hybridoma technology for monoclonal antibody production
1982	FDA approval of recombinant insulin—Birth of Biopharmaceutical Industry
1983	Invention of Polymerase Chain Reaction (PCR)
1983	First genetic mapping of disease, Huntington's disease
1986	First FDA approval of therapeutic monoclonal antibody
1989	Discovery of CFTR Gene
1997	First FDA approval for therapeutic monoclonal antibody for oncology
1998	First co-approval of diagnostic alongside therapeutic
2000	First draft of Human Genome Draft
2001	FDA approval of first wave of tyrosine kinase inhibitors for oncology
2012	FDA approval for CFTR using corrective function based on mechanism
2014	First FDA approval of checkpoint inhibitors
2017	First CAR-T Therapy approval
2023	FDA approval of first CRISPR-based therapy

therapeutic monoclonal antibodies remain a powerful force alongside novel approaches with emerging methods. As the number of targets broadens, so does the need for refinement in our molecular tools for diagnostics. As seen with the case of cystic fibrosis, a monogenic disease quickly becomes seemingly complex as additional genetic information becomes readily available. There are hundreds of disease-causing mutations and at least four targeted therapies developed based on restoring protein function; yet, there is still a gap between identifying a new mutation and determining the best therapeutic. As more multifactorial approaches are taken to combat various cancers and drug resistance, the complementing biomarkers to understand cell state dynamics will be ever more critical.

References

1. Woese, C. R. (2004). A new biology for a new century. *Microbiology and Molecular Biology Reviews, 68*(2), 173–186.
2. Giani, A. M., Gallo, G. R., Gianfranceschi, L., & Formenti, G. (2019). Long walk to genomics: History and current approaches to genome sequencing and assembly. *Computational and Structural Biotechnology Journal, 18*, 9–19. https://doi.org/10.1016/j.csbj.2019.11.002
3. Morell, V. (1997). Microbial biology: Microbiology's scarred revolutionary. *Science, 276*, 699–702. https://doi.org/10.1126/science.276.5313.699
4. Shah, E. (2016). A tale of two biographies: The myth and truth of Barbara McClintock. *History and Philosophy of the Life Sciences, 38*(4), 1–12. JSTOR.
5. Henderson, D. A., & Moss, B. (1999). Smallpox and vaccinia. In S. A. Plotkin & W. A. Orenstein (Eds.), *Vaccines* (3rd ed.). Saunders. Chapter 6.
6. Karamanou, M., Protogerou, A., Tsoucalas, G., Androutsos, G., & Poulakou-Rebelakou, E. (2016). Milestones in the history of diabetes mellitus: The main contributors. *World Journal of Diabetes, 7*(1), 1–7.
7. Riedel, S. (2005). Edward Jenner and the history of smallpox and vaccination. *Proceedings (Bayl Univ Med Cent)., 18*(1), 21–25.
8. Boylston, A. (2019). The origins of inoculation. J R Soc Med. 2012 Jul;105(7):309-13 e sequencing and assembly. *Computational and Structural Biotechnology Journal, 18*, 9–19.
9. Peebles, J. M. (1900). *Vaccination a cure and a menace to personal liberty, with statistics showing its dangers and criminality*. The Temple of Health Publishing Company.
10. Esparza, J., Ledeman, S., Nitsche, A., & Damasca, C. R. (2020). Early smallpox vaccine manufacturing in the United States: Introduction of the "animal vaccine" in 1870, establishment of "vaccine farms", and the beginnings of the vaccine industry. *Vaccine, 38*(30), 4773–4779, ISSN 0264-410X.
11. Tyagi, U., & Barwal, K. C. (2020). Ignac Semmelweis-father of hand hygiene. *The Indian Journal of Surgery, 82*(3), 276–277.
12. Hutt, P. B. (2018). The evolution of federal regulation of human drugs in the United States: An historical essay. *American Journal of Law & Medicine, 44*(2–3), 403–451.
13. Vecchio, I., Tornali, C., Bragazzi, N. L., & Martini, M. (2018). The discovery of insulin: An important milestone in the history of medicine. *Frontiers in Endocrinology (Lausanne), 9*, 613.
14. Lewis, G. F., & Brubaker, P. L. (2021). The discovery of insulin revisited: Lessons for the modern era. *The Journal of Clinical Investigation, 131*(1), e142239.
15. Rosenfeld, L. (2002). Insulin: Discovery and controversy. *Clinical Chemistry, 48*(12), 2270–2288. PMID: 12446492.
16. Cassier, M., & Sinding, C. (2008). 'Patenting in the public interest:' Administration of insulin patents by the University of Toronto. *History and Technology, 24*(2), 153–171.
17. Hirsch, I., Juneja, R., Beals, J. M., Antalis, C. J., & Wright, E. E. (2020). The evolution of insulin and how it informs therapy and treatment choices. *Endocrine Reviews, 41*(5), 733–755.
18. Kahn, C. R., & Roth, J. (2004). Berson, Yalow, and the JCI: The agony and the ecstasy. *The Journal of Clinical Investigation, 114*(8), 1051–1054.
19. Wu, A. H. B. (2006). A selected history and future of immunoassay development and applications in clinical chemistry. *Clinica Chimica Acta, 369*(2), 119–124. ISSN 0009-8981.
20. Mitra, S., & Tomar, P. C. (2021). Hybridoma technology; advancements, clinical significance, and future aspects. *Journal, Genetic Engineering & Biotechnology, 19*(1), 159.

21. Chial, H. (2008). Rare genetic disorders: Learning about genetic disease through gene mapping, SNPs, and microarray data. *Nature Education, 1*(1), 192.
22. Farrell, P. M., Rock, M. J., & Baker, M. W. (2020). The impact of the CFTR gene discovery on cystic fibrosis diagnosis, counseling, and preventive therapy. *Genes (Basel), 11*(4), 401.
23. Foil, K. E., Powers, A., Raraigh, K. S., Wallis, K., Southern, K. W., & Salinas, D. (2019). The increasing challenge of genetic counseling for cystic fibrosis. *Journal of Cystic Fibrosis, 18*(2), 167–174.
24. Roda, J., Pinto-Silva, C., Silva, I. A. I., Maia, C., Almeida, S., Ferreira, R., & Oliveira, G. (2022). New drugs in cystic fibrosis: What has changed in the last decade? *Therapeutic Advances in Chronic Disease, 13*, 20406223221098136.
25. Jaques, R., Shakeel, A., & Hoyle, C. (2020). Novel therapeutic approaches for the management of cystic fibrosis. *Multidisciplinary Respiratory Medicine, 15*(1), 690.
26. Luca, F., Salvatore, S., & Massimo, L. (2018). Evolution of cancer pharmacological treatments at the turn of the third millennium. *Frontiers in Pharmacology, 9*, 1300.
27. Goldman, J. M., & Daley, G. Q. (2007). Chronic myeloid Leukemia — A brief history. In *Myeloproliferative disorders. Hematologic malignancies*. Springer.
28. Lu, R. M., Hwang, Y. C., Liu, I. J., et al. (2020). Development of therapeutic antibodies for the treatment of diseases. *Journal of Biomedical Science, 27*, 1.
29. Jørgensen, J. T., Winther, H., Askaa, J., Andresen, L., Olsen, D., & Mollerup, J. (2021). A companion diagnostic with significant clinical impact in treatment of breast and gastric cancer. *Frontiers in Oncology, 11*, 676939. https://doi.org/10.3389/fonc.2021.676939
30. Min, H. Y., & Lee, H. Y. (2022). Molecular targeted therapy for anticancer treatment. *Experimental & Molecular Medicine, 54*, 1670–1694. https://doi.org/10.1038/s12276-022-00864-3
31. Chen, Y. J., Abila, B., & Mostafa, K. Y. (2023). CAR-T: What is next? *Cancers (Basel), 15*(3), 663.
32. Hahn, W. C., Bader, J. S., Braun, T. P., Califano, A., Clemons, P. A., Druker, B. J., Ewald, A. J., Fu, H., Jagu, S., Kemp, C. J., Kim, W., Kuo, C. J., McManus, M., Mills, B., Mo, X., Sahni, N., Schreiber, S. L., Talamas, J. A., Tamayo, P., Tyner, J. W., Wagner, B. K., Weiss, W. A., & Gerhard, D. S. (2021). Cancer target discovery and development network. An expanded universe of cancer targets. *Cell, 184*(5), 1142–1155.

Different Career Paths in the Biopharma Industry

Exploring Marketing and Product Management

Leena Paul

Contents

My Journey

My role at Agilent is as a Product Specialist, a journey that began after completing my Ph.D. in Biochemistry from the University of Maryland in 2000. Balancing family commitments, I spent three years at home with my children before pursuing my interest in business and science through a part-time MBA program at Johns Hopkins University's Shade Grove location. Realizing my passion for the business side of the biopharma industry, I worked as a marketing specialist in my initial years, crafting documents for nonscientific audiences.

My first full-time position was a sales representative role at Integrated DNA Technologies, a genomics solutions provider. I faced challenges adapting from a science background to sales but exceeded quotas, quadrupling them in the second year. Four years later, I joined Amaxa (now a division of Lonza, a major manufacturer of healthcare research products) as a Technical Sales Account Manager. I thrived in a challenging market, building confidence in my career trajectory. Following Amaxa's acquisition by Lonza, I embraced new roles as a Key Account Manager and Distribution Channel Manager, expanding my responsibilities to cover Canada. My work involved liaising with stakeholders

L. Paul (✉)
Agilent Technologies, Santa Clara, CA, USA
e-mail: Leena.paul@agilent.com

S. Taylor, M. Campbell (eds.), *Women in Biopharma*, Women in Engineering and Science, https://doi.org/10.1007/978-3-031-80348-2_2

from CEOs to benchtop researchers, addressing technical and nontechnical issues, and supporting clients.

Seeking new opportunities to balance family commitments and career growth, I joined Agilent as a Product Specialist, covering a geographic area with minimal travel demands and specializing in Seahorse XF Technology (instruments that measure real-time analysis of cellular energy metabolism). This technology lets scientists to see how cells produce and use energy in real-time. Think of it like a fitness tracker for cells! Just as a fitness tracker helps you understand how your body is performing during exercise, the Seahorse XF Analyzer helps scientists understand how cells are performing their "workouts" inside our bodies. With the Seahorse XF Analyzer, scientists can uncover new insights into diseases like cancer, diabetes, and neurodegenerative disorders. They can also develop better treatments and therapies to help patients live healthier lives.

I expanded into other technologies and took on mentoring roles, gaining insights into human nature and skill sets through involvement in the hiring process. Flexibility in sales allowed me to manage family obligations while staying focused on customer satisfaction and revenue generation. Embracing changes in the dynamic market space, particularly during and after COVID-19, and serving as a role model for juniors contributed to my success, culminating in winning the President's Club Award in 2023. Acknowledged by the Agilent CEO and President for resilience, dedication, and excellent customer support, I am proud to have contributed to Agilent's long-term growth and success. While my current role involves collaborating with multiple Account Managers, my future trajectory at Agilent entails transitioning into sales management positions or exploring opportunities within other corporate business roles.

1 Scientists

Scientists play a pivotal role in the biopharmaceutical industry, driving innovation, and spearheading the development of new drugs, therapies, and biologics. These individuals possess expertise in fields such as molecular biology, biochemistry, genetics, and pharmacology. They are responsible for designing and conducting experiments, analyzing data, and interpreting results. Whether they are exploring new drug candidates, developing innovative therapies, or investigating the mechanisms of disease, scientists and researchers are the driving force behind scientific progress in the biotech industry.

Here's a closer look at their roles, responsibilities, and contributions: Scientists are at the forefront of research and discovery efforts within biopharma companies. They design and execute experiments, conduct literature reviews, and explore novel scientific concepts to identify potential targets for drug development. By leveraging their expertise in fields such as molecular biology, biochemistry, pharmacology, and immunology, they uncover new insights into disease mechanisms and pathways, laying the groundwork for the development of innovative therapies.

Once potential drug targets are identified, scientists embark on the process of drug development and optimization. They design and test candidate molecules, assess their efficacy and safety profiles, and optimize their pharmacological properties to enhance therapeutic outcomes. Through iterative experimentation and data analysis, they refine drug candidates to maximize their clinical potential while minimizing risks and side effects.

Preclinical and translational research are critical stages in the drug development process, bridging the gap between laboratory discovery and clinical testing. Scientists and researchers conduct preclinical studies using cell culture models, animal models, and in vitro assays to evaluate the efficacy, toxicity, and pharmacokinetics of potential drug candidates. They generate robust preclinical data to support Investigational New Drug (IND) applications and advance promising candidates into clinical trials.

Scientists collaborate closely with clinical teams to design and implement clinical trials that evaluate the safety and efficacy of investigational drugs in human subjects. They contribute their scientific expertise to protocol development, patient selection criteria, and endpoint selection, ensuring that the trials are scientifically rigorous and ethically sound. Throughout the clinical trial process, they monitor data quality, analyze results, and interpret findings to inform decision-making and regulatory submissions.

Scientists also participate in regulatory compliance and documentation to ensure compliance with regulatory requirements and guidelines throughout the drug development process. They prepare and review regulatory documents, including IND applications, investigator's brochures, and clinical study reports, to support regulatory submissions and approvals. By maintaining meticulous documentation and adhering to good laboratory practice (GLP) and good clinical practice (GCP) standards, they facilitate the regulatory review process and expedite the path to market for new therapies.

Collaboration is fundamental to the success of Scientists in a biopharma company. They collaborate with cross-functional teams, including chemists, biostatisticians, clinicians, and regulatory affairs specialists, to leverage diverse perspectives and expertise in advancing drug development programs. Effective communication skills are essential for scientists to convey complex scientific concepts, present data, and share insights with colleagues, stakeholders, and external partners.

2 Sales, Marketing, and Business Development

In a biopharma company, the roles of sales, marketing, and business development are crucial for driving growth, increasing market share, and ensuring the successful commercialization of products and services.

2.1 Sales

Sales representatives play a vital role in building relationships with healthcare providers, physicians, hospitals, pharmacies, and other clients. They engage in direct sales activities, such as product demonstrations, presentations, and negotiations, to promote their company's products and services.

Sales personnel identify and pursue opportunities to expand their company's market presence and reach new customer segments. They conduct market research, assess customer needs, and develop targeted sales strategies to capitalize on emerging trends and opportunities. Beyond initial sales, sales personnel provide ongoing support to clients, addressing inquiries, resolving issues, and ensuring customer satisfaction. By maintaining strong relationships and delivering value-added services, they foster customer loyalty and retention, driving repeat business and revenue growth.

Inside sales and outside sales personnel play important roles in driving business growth and customer satisfaction, but they operate in different environments and focus on distinct aspects of the sales process.

Inside sales personnel operate from a more centralized location, primarily engaging with clients through phone calls, emails, and virtual meetings. They focus on lead generation, prospecting, and customer support. They proactively reach out to potential clients to discuss product offerings, address inquiries, and facilitate order processing. Inside sales representatives play a crucial role in generating and qualifying leads, providing sales support to field representatives, and building relationships with clients through effective communication and personalized engagement. Their efforts contribute to driving customer satisfaction, retention, and revenue growth.

In contrast, outside sales personnel in this industry are often comprised of field Account Managers and Product Specialists. The Account Managers are called the "hunters" and could be nontechnical, whereas Product Specialists are more technically advanced and provide solutions to clients. Both meet with clients, travel to customer sites, laboratories, and research facilities to conduct in-person meetings and presentations. They engage in consultative selling, working closely with clients to understand their unique needs, challenges, and objectives, and they provide tailored solutions and recommendations. Outside sales personnel focus on expanding market presence, driving revenue growth, and nurturing long-term relationships with key clients through face-to-face interactions and strategic account management.

2.2 Marketing

The marketing team plays a critical role in biopharma companies, shaping perceptions, driving demand, and facilitating the successful commercialization of products and services. Here are some key marketing roles within a biopharma company:

Product Management Product managers oversee the development, positioning, and promotion of specific products or product lines. They conduct market research to identify customer needs, develop product strategies, and collaborate with cross-functional teams to ensure product success throughout its lifecycle. Product launch managers oversee the planning and execution of new product launches, coordinating cross-functional teams to ensure a successful market entry. They develop launch strategies, create launch plans, and execute marketing activities to generate awareness, excitement, and adoption of new products in the market. A successful launch of any product needs research on market trends, customer preferences, and competitor activities. They use various research methodologies to provide insights that inform marketing strategies, product development decisions, and market positioning.

Digital Marketing Digital marketers leverage online channels such as websites, social media, email marketing, and search engine optimization (SEO) to promote products, engage with clients, and generate leads. They develop digital marketing campaigns, track performance metrics, and optimize strategies to enhance brand visibility and drive sales.

Content Marketing Content marketers create valuable, relevant content, such as blog posts, whitepapers, case studies, and videos, to educate, inform, and engage target audiences. They develop content strategies aligned with marketing objectives, conduct keyword research, and collaborate with subject matter experts to produce high-quality content that resonates with clients.

Customer Relationship Management (CRM) CRM specialists manage customer databases, implement CRM systems, and analyze customer interactions to improve customer engagement and retention. They use CRM data to segment clients, personalize marketing communications, and track customer interactions throughout the sales cycle.

Business/Market Development While marketing focuses on promoting and positioning products or services in the market, business development involves identifying and pursuing growth opportunities, partnerships, and strategic alliances to expand the company's reach and revenue streams.

Business development professionals scout for emerging technologies, therapeutic areas, and market trends that align with the company's strategic objectives. Through rigorous market research and analysis, they assess the potential value and feasibility of numerous opportunities, including licensing agreements, collaborations, and acquisitions. Additionally, business development teams cultivate and nurture relationships with external stakeholders, including biotech startups, academic institutions, and pharmaceutical companies. They explore potential partnerships and collaborations that drive value creation and enhance the company's competitive positioning.

Business development executives lead negotiations and deal-making processes, leveraging their expertise in deal structuring, negotiation strategies, and industry dynamics to secure favorable terms and drive mutually beneficial outcomes. They collaborate closely with cross-functional teams, including research and development, marketing, and regulatory affairs, to develop and execute commercialization strategies for new products or technologies. By identifying opportunities for geographic expansion, market entry, and portfolio optimization, business development professionals contribute to the company's overall growth and success in the dynamic and competitive biopharma landscape.

Kelle Hammock, Agilent Global Market Development Manager & Clinical Partnerships and Collaborations, shares her valuable insight: "Our Genomics Innovation Group (GIG) for the Integrated Genomics Division (IGD), under Kevin Meldrum, VP/GM, has two sectors; the first group focuses on emerging technologies and works closely with Early-Stage Partnerships. The second group focuses on collaborations with Academic/Cancer Centers or industry partners, such as biopharma, using commercially available products. My role on the GIG team and in Market Development centers on the second sector, identifying and streamlining Collaborations/Partnerships, as well as supporting the Original Equipment Manufacturer (OEM) business.

This role is extremely vital when initiating partnership discussions, within any of the above-mentioned market segments, using Agilent's commercialized IGD products. We work closely with the Product Management team, R&D, Field Application Scientist, Sales Leadership, Technical Support, Regulatory/Quality Assurance, and other key stakeholders throughout the three steps of a partnership, including ideation and strategic decision-making, brainstorming and negotiation phase, followed by commercialization and execution. It is critical that the core team works together and in tandem, with the end goal of successfully enabling and executing diverse partnerships/collaborations that synergistically benefit Agilent and our partners by addressing key unmet needs, while also allowing Agilent to expand into new or existing markets, without having to build it ourselves."

3 Sales Management

Sales management refers to the process of planning, implementing, and controlling sales activities within an organization. It involves overseeing the sales team, setting sales objectives, devising strategies to achieve those objectives, and monitoring the performance of the sales force.

Successful sales managers possess a multifaceted skill set, encompassing leadership, communication, sales expertise, strategic thinking, coaching, problem-solving, data analysis, customer focus, adaptability, and emotional intelligence. They provide direction and motivation to their teams, communicate effectively with both internal stakeholders and clients, and leverage their deep understanding of sales processes and best practices to drive revenue growth. Sales managers excel in strategic planning, coaching and developing their teams, analyzing sales data to make informed decisions, and prioritizing customer satisfaction. Their ability to adapt to changing market dynamics, solve problems, and build strong relationships with both team members and clients is key to their success in achieving sales targets and driving business growth.

4 Scientific Support and Field Application Scientists

Scientific Support and Field Application Scientists play critical roles at Agilent, providing expertise and assistance to clients in utilizing their company's products and solutions effectively.

Technical Support Technical support specialists serve as the frontline support for clients, offering assistance with troubleshooting, product inquiries, and resolving technical issues. They provide timely and accurate responses to customer inquiries via phone, email, or online platforms. Technical support professionals possess in-depth knowledge of the product portfolio and technical specifications, allowing them to diagnose problems, offer solutions, and escalate complex issues as needed. They play a crucial role in maintaining customer satisfaction by ensuring that issues are resolved promptly and effectively, thus helping clients maximize the performance and value of their products. If you are technically strong, but don't want to face clients face-to-face, this is a perfect role for you.

Agilent Support Engineering Manager Joyce Velez says, "My position is Remote Support Engineering (RSE) Manager. My team is a mix of former scientists that specialize in helping scientists and technicians world-wide. Their support covers anything related to hardware, software, and applications. Our portfolio includes instrumentation to perform scientific research, drug manufacturing, in-vitro diagnostics, etc. It includes imaging systems, real time cellular analyzers, flow cytometers, plate readers, and liquid handlers, to name a few. They are located across the US and Europe. While my team's focus is to as-sist clients, my focus is to enable, protect, and support my team.

Before I joined Agilent, I was a senior scientist at a private research institute. My educational formation started with a bachelor's in science, followed by a PhD in toxicology, and post-doctoral research in neurotoxicity and neurology. I started at Agilent as a technical support scientist, with a very small group that eventually grew into a global contact center with 15 remote service engineers split be-tween the US and Europe. Challenges are mostly related to time zones because my team supports clients live for 18 hours straight. That is also one of my favorite parts of the job making it more interesting and learning about how different the same support can happen in different regions of the world. It provides me with continuous interaction with people from diverse backgrounds and cultures. My short-term career goal is to continue to build a more robust and prepared remote support team with more coverage across the globe. In the long term, I plan to continue to grow in Agilent within our customer-focused organization."

Field Application Scientists Field Application Scientists (FAS) are experts in specific scientific applications and technologies. They work closely with clients to understand their research needs, provide technical guidance, and demonstrate the capabilities of instruments and solutions in real-world settings. FAS professionals collaborate with sales teams during presales consultations, helping clients select the right products and configure solutions tailored to their requirements. Additionally, they conduct product demonstrations, training sessions, and workshops to educate clients on best practices and to maximize the utility of various products and solutions in their research workflows. By serving as trusted advisors and technical experts, Field Application Scientists play a pivotal role in building strong customer relationships, driving the adoption of products, and contributing to overall customer success.

5 Nontechnical Positions in a Biopharma Company

In biopharma companies, nontechnical positions are essential for various aspects of operations, administration, and business functions. These roles contribute to the overall success of the company by providing support in areas such as administration, finance, human resources, regulatory affairs, marketing, sales, and project management. Below are some common nontechnical positions found in biopharma companies:

Administration Administrative assistants, office managers, and receptionists play vital roles in maintaining smooth day-to-day operations, managing administrative tasks, scheduling meetings, coordinating events, and handling correspondence.

Finance Finance professionals, including accountants, financial analysts, and controllers, manage financial planning, budgeting, forecasting, and reporting activities. They ensure compliance with regulatory requirements, analyze financial data, and provide strategic financial guidance to support business decisions.

Human Resources Human resource (HR) professionals oversee recruitment, employee relations, performance management, training and development, and compensation and benefits administration. They play a critical role in fostering a positive work culture, managing talent, and ensuring compliance with labor laws and regulations.

Regulatory Affairs Regulatory affairs specialists are responsible for ensuring compliance with regulatory requirements and obtaining approvals for the development, manufacturing, and marketing of pharmaceutical products. They liaise with regulatory agencies, prepare regulatory submissions, and monitor changes in regulations to ensure product compliance.

Project Management Project managers oversee the planning, execution, and monitoring of projects related to product development, clinical trials, manufacturing, and commercialization. They coordinate cross-functional teams, manage project timelines and budgets, and ensure that project objectives are met within scope and quality standards.

Quality Assurance and Compliance Quality assurance (QA) and compliance professionals ensure that biopharma products meet quality standards, regulatory requirements, and industry best practices. They develop and implement quality management systems, conduct audits, and provide oversight to manufacturing processes to maintain product quality and safety.

Supply Chain and Logistics Supply chain and logistics professionals manage the procurement, inventory management, and distribution of raw materials, components, and finished products. They optimize supply chain processes, minimize costs, and ensure the timely delivery of products to clients while maintaining compliance with regulatory requirements.

Legal and Intellectual Property Legal and intellectual property (IP) professionals provide legal counsel, contract negotiation, and IP protection services to biopharma companies. They draft and review contracts, manage IP portfolios, and provide guidance on legal and regulatory matters to support business activities and protect company assets.

Corporate Communications Corporate communications specialists manage internal and external communication strategies, including media relations, public relations, investor relations, and corporate branding. They develop communication plans, handle crisis communications, and promote the company's image and reputation to stakeholders.

6 Conclusion

As you embark on the next chapter of your life, you may find yourself faced with a myriad of career choices, each offering its own unique opportunities and possibilities. Depending on your education and your drive to learn, the choices are limitless.

The landscape of a research biopharma company is rich and diverse, comprising a multitude of professions that collectively contribute to the advancement of science and the development of life-saving therapies. From scientists and researchers driving discovery in the laboratory to regulatory affairs specialists navigating the complexities of compliance, each profession plays a unique and vital role in the biotech ecosystem. By harnessing the collective expertise and ingenuity of these professionals, research biotech companies continue to push the boundaries of innovation and improve the quality of life for people around the world.

For young professionals embarking on their career journey, my advice is simple yet profound: follow your passion and stay attuned to market trends. It's crucial to pursue opportunities that align with your interests, values, and aspirations. By following your heart and pursuing work that ignites your passion, you'll find fulfillment and satisfaction in your career path.

Additionally, staying informed about market trends and industry developments is essential for career success. Be cognizant of emerging technologies, evolving market demands, and shifting consumer preferences. This knowledge will empower you to make informed decisions, seize opportunities, and adapt to changing landscapes.

Prioritize finding the right cultural fit within organizations. A positive workplace culture that values diversity, collaboration, and innovation fosters a supportive environment where you can thrive and grow professionally. Seek out organizations that align with your values and provide opportunities for personal and professional development.

Remember, career advancement is not just about climbing the corporate ladder; it's about finding fulfillment, making meaningful contributions, and continuously learning and growing. By following your heart, staying informed, and cultivating the right cultural fit, you'll pave the way for a fulfilling and successful career journey.

The Mysteries of a Biopharma Legal Department

Jasbir Sagoo, Lata Kirtivasan, Leslie Stevens, Miranda Pierce, Shannon Jamison, Tsering Ringu, and Virginia MacSuibhne

Contents

J. Sagoo (✉)
Intellectual Property – Agilent Technologies, Boston, MA, USA
e-mail: jasbir.sagoo@agilent.com

L. Kirtivasan
Contracts Team – Agilent Technologies, Manesar, Gurgaon, Haryana, India
e-mail: lata_kirtivasan@agilent.com

L. Stevens
Compliance Department – Agilent Technologies, Santa Clara, CA, USA
e-mail: leslie.stevens@agilent.com

M. Pierce
Contracts Team – Agilent Technologies, Tallahassee, FL, USA
e-mail: Miranda_Pierce@agilent.com

S. Jamison
Karl Storz, El Segundo, USA
e-mail: shannon.jamison@karlstorz.com

T. Ringu
Contracts Team – Agilent Technologies, Cedar Creek, TX, USA
e-mail: Tsering_ringu@agilent.com

V. MacSuibhne
Agilent Technologies, Santa Clara, CA, USA

San Jose, USA

S. Taylor, M. Campbell (eds.), *Women in Biopharma*, Women in Engineering and Science, https://doi.org/10.1007/978-3-031-80348-2_3

1 Introduction

Who Is in the Legal Department of a Biopharma Company, How Did They Get There, and What do They Do All Day?

A legal department plays a critical role in a biopharma company, from protecting the intellectual property behind the science to ensuring contracts cover key elements for collaboration and sales, and developing policies and procedures so employees

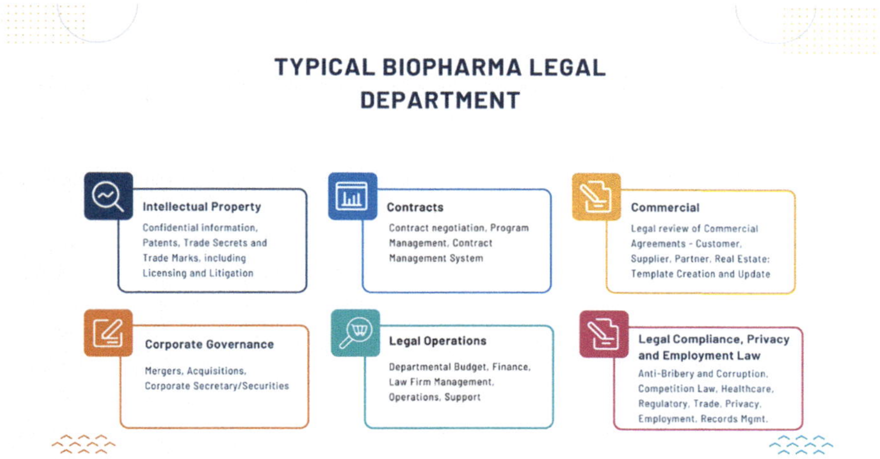

Fig. 1 Typical biopharma legal department

know how to operate within the boundaries of the law. This chapter explores the different functions, subjects, and teams behind a legal department in biopharma, including the path that was taken to get here. First, let's look at the typical functional department setup of a legal team in biopharma (Fig. 1).

2 Intellectual Property (IP)

Jasbir Sagoo, Associate Vice President (AVP) of Intellectual Property (IP)—Agilent Technologies, Inc.

2.1 What Is Intellectual Property?

Intellectual Property (IP) law deals with property rights of intangible creations of the human mind. Think of a formula or process created to develop a drug or diagnostic device. IP law protects and enforces the rights of the creators, inventors, and owners of inventions for things like writings, music, designs, and other works. There are several areas of IP, including patents, trademarks, copyrights, confidential information, know-how, and trade secrets.

2.2 What Do IP Practitioners Do?

The three broadest areas of IP practice involve counseling, protecting, and enforcing. Depending on whether one chooses to be in private practice in a law firm or in-house in a biotech/biopharma company, the "client" is either an inventor/company that comes to the law firm for its services, or the scientists/engineers employed by the company, respectively.

Client counseling involves how best to protect the IP that the client has or is developing. IP protection involves obtaining a granted patent or registering a trademark or copyright to protect the client's asset. In the case of a patent or trademark, the process involves preparing and filing an application with the Patent and Trademark Offices (PTO) in various countries and responding to actions issued by the PTO until the patent issues or the trademark is registered. Enforcement of IP involves protecting the owner of the IP against infringing uses. This may involve litigation in the court system.

2.3 What Do I Need to Study to Become an IP Practitioner?

For patents, becoming a patent agent or patent attorney blends both technical expertise and legal knowledge. The ability to understand the idea and the underlying technology and translate it into a patent application is key. The IP practitioner also needs to write and communicate clearly to advocate for their clients when working with patent examiners at the PTO to secure a patent. A technical background in science (biology, chemistry, physics, bioengineering, engineering) is required, with a minimum of a Bachelor of Science degree. Depending on the country, an individual can become a patent agent by passing a patent bar exam and/or a patent attorney by also completing a law degree.

For trademarks and copyrights, a law degree is required, but a technical background or degree is not essential. Of note, General Attorneys without a technical background or degree can support the company's IP assets by drafting IP-related contracts or licensing agreements.

3 Different Specialties Within IP

There are two general categories of IP: those that are seen by the public because they are registered at PTOs (Patents, Trademarks, Copyrights), and those that are not seen by the public (confidential information, know-how, and trade secrets) (Fig. 2).

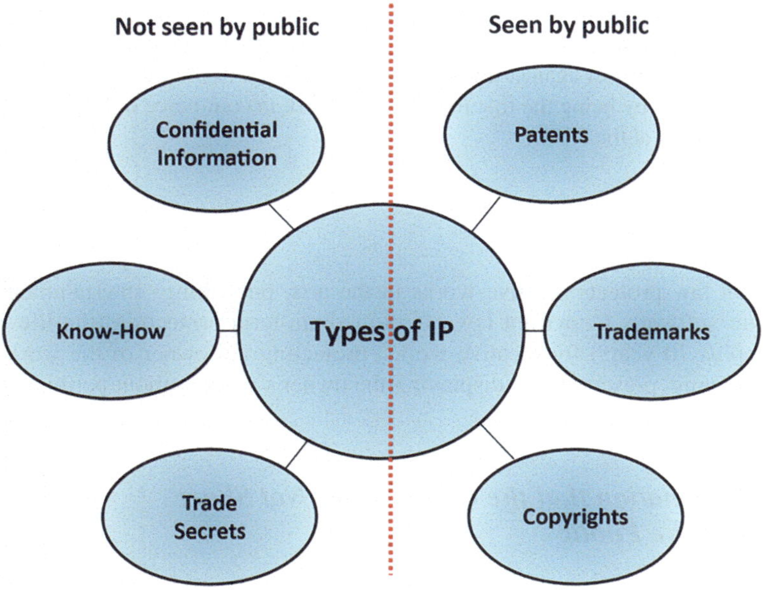

Fig. 2 Different types of Intellectual Property

3.1 Information Shared with the Public by the Owner Notifying the World of the Protections

3.1.1 Patents

Patent law grants short-term protection for new inventions, which can be products, processes, or designs. Patent law promotes the sharing of new developments with others to foster innovation by providing 20 years of protection, after which time the invention becomes available for all to use. During the life of the patent, prior to its expiration, the patent owner has the right to exclude others from making, using, selling, offering to sell, or importing the patent-protected invention, although they can license or assign the patent to others. Patents are "territorial," meaning they are unique to a particular country and only offer protection in that specific country. As such, many large biopharma companies file patents in many countries, while smaller biopharma companies file patents in major markets or where they conduct most of their business, such as the United States, Europe, China, and Japan.

3.1.2 Trademarks

Trademark law protects a word, phrase, symbol, or design that is used by an entity to identify the source or provider of goods or services. Trademarks offer potentially perpetual protection (but require renewal every 10 years). Examples include Agilent's spark logo, Apple's iconic apple logo, and Nike's widely recognized swoosh sign. Trademark owners can prevent others from using identical or confusingly similar marks that might otherwise result in consumers misidentifying the true source of a product or service. Trademark laws protect against infringement and dilution of the mark. Rights to trademarks are gained by being the first to use a trademark in commerce or being the first to register the mark at the PTO.

3.1.3 Copyrights

Copyright law protects creative works in the arts, publishing, entertainment, and computer software. Copyright law offers medium-term protection (the life of the creator plus 70 years) for creative works, protecting the owner of the work from others copying, presenting, or displaying the owner's work without permission.

3.2 Information that the Owner Does Not Share with the Public

3.2.1 Confidential Information

Confidential information includes internal proprietary information, i.e., anything unique to the company that is material to business operations and cannot be learned externally. It may include documents or materials, physical areas such as a manufacturing site layout, procedures, processes, marketing plans, payroll/personnel records, computer programs, and virtually anything else a business maintains as not publicly disclosed. It can exist in many forms, including written, spoken, observed, or electronically stored information.

3.2.2 Trade Secrets

Trade secrets encompass a particular type of confidential information that derives substantial economic value from remaining undisclosed outside of the company or not being widely known; for example, the secret formula of Coca-Cola. In a biopharma company, trade secrets can include chemical formulations, manufacturing processes, and computer algorithms. Trade secrets offer potentially perpetual protection as long as the knowledge is kept secret.

All trade secrets are confidential information, but not all confidential information rises to the level of a trade secret. Trade secrets are afforded the highest level of protection of all proprietary information and require the biopharma company to take reasonable steps to protect the trade secrets by implementing measures to limit access to the trade secrets and sharing them with only those employees who need to know such information to do their jobs.

3.2.3 Know-How

Know-how refers to technical skills, including information, knowledge, techniques, and experiences—essentially what an individual acquires as they become experienced or specialists in a particular field within a biopharma company. It comprises knowledge obtained while on the job, including how not to do something as well as how to do something well. Know-how may be associated with trade secrets, confidential, or proprietary information. While know-how may include trade secrets, it generally does not require the same type and level of protection as trade secrets. Figure 3 shows how confidential information, know-how, and trade secrets fit together.

In the biopharma space, there is an interplay between all these different types of IP, as illustrated by Agilent's xCELLigence RTCA eSight™ product (Fig. 4).

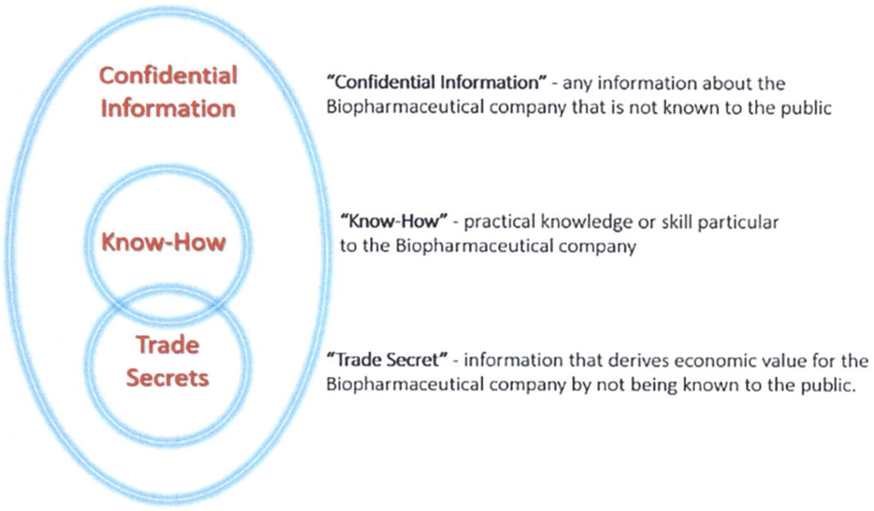

Fig. 3 Relationship between confidential information, know-how and trade-secrets

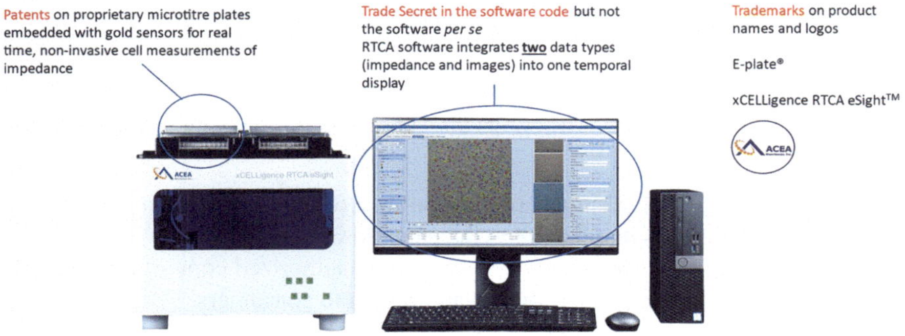

Fig. 4 Different IP types with product example

4 Working with IP in a Biopharma Company

Attorneys, contract managers, paralegals, and other corporate professionals frequently work with contracts and licensing agreements that contain IP terms and considerations.

4.1 IP-Related Contracts

IP owners may commercialize products that embody or contain IP in many ways. Some manufacture, market, sell, and distribute directly, while others perform some of these activities and engage third parties to do the rest. Some IP owners contract exclusively with third parties to carry out these activities on their behalf. For commercialization of IP, it is essential for both the IP owner and the third party to enter into an agreement that sets forth the IP rights granted and the obligations of each party. Examples of IP-related agreements include Material Transfer Agreements (MTAs), Confidentiality/Non-Disclosure Agreements (CDAs/NDAs), Commercialization, Distribution and Supply Agreements, Collaboration Agreements, Software License and Maintenance Agreements, Outsourcing Agreements, and Licensing Agreements.

4.2 IP Licensing

IP owners may generate revenue through licensing of their IP to third parties in exchange for a licensing fee and/or royalty. An IP license authorizes third parties to conduct activities that would otherwise infringe the IP owner's rights. The IP owner determines the scope and terms of the license, such as the field of use, duration of the license, and geography.

4.3 IP Infringement and Litigation

IP infringement occurs when a third-party infringer uses the IP of another without permission. IP infringement is a serious threat to a biopharma company's livelihood and includes:

Patent Infringement: An infringer creates a product or service based on innovations, processes, or designs that belong to the patent owner.

Copyright Infringement: An infringer makes an unauthorized copy of a copyright-protected piece of literature or music and sells it to consumers.

Trademark Infringement: An infringer attempts to sell its product using an unauthorized logo or related mark that is confusingly similar to the trademark owner's.

Trade Secret Misappropriation: An infringer takes and uses trade secrets of the IP owner without permission.

4.4 IP Litigation in the United States

The IP Owner ("plaintiff") can bring a lawsuit against a third-party infringer ("defendant") who improperly uses its IP in civil court. The IP owner may seek to prevent the infringer from selling products that use the IP without permission and seek monetary damages from the infringer.

4.4.1 Patent Litigation

In patent litigation, the plaintiff alleges either (i) direct infringement, where the defendant has directly manufactured, used, sold, and/or imported the plaintiff's patented invention, method, or service without permission; or (ii) indirect infringement, where the defendant enabled or induced others to commit the infringement.

Patent litigation takes an average of 3–5 years, with an average cost of about \$4–\$6 million (Patent Litigation Statistics: Feuds about Intellectual Property, legaljobs.io). Cases are usually tried before a jury.

4.4.2 Copyright Litigation

In copyright litigation, a plaintiff seeks to prevent the defendant's unauthorized use of the copyrighted materials and to recover damages. The plaintiff must establish that it owns a valid copyright (registered with the Copyright Office) and that the defendant used the copyrighted information without appropriate permission.

Copyright litigation may take over a year to litigate with an average cost of about \$280,000 (Copyright litigation 101, thomsonreuters.com).

4.4.3 Trademark Litigation

In trademark litigation, a plaintiff must establish a likelihood of confusion between the plaintiff's trademark and the defendant's use of a mark or logo, which may cause customers to think the plaintiff is providing the products or services when in fact they are not. A likelihood of confusion may be established by a plaintiff by demonstrating that the defendant's mark or logo is similar in image or name and reduces the plaintiff's trademark's distinctiveness, thereby diluting its value.

Trademark infringement lawsuits that advance to trial take over a year to litigate and cost anywhere from $375,000 to $2 million (https://securityboulevard.com/2023/05/what-is-trademark-infringement-and-how-costly-is-it-to-your-business/) (Trademark litigation 101: Legal Blog, thomsonreuters.com).

4.4.4 Trade Secret Litigation

In trade secret litigation, the protected information at issue has three elements (Trade secret litigation 101, thomsonreuters.com):

(i) It must have "actual or potential economic value" by virtue of not being known or readily ascertainable by others.
(ii) It provides a "competitive advantage" to the biopharma company.
(iii) The information must be subject to "reasonable efforts" to maintain its secrecy.

The plaintiff must prove in court that the trade secret was "misappropriated or wrongfully taken" by the defendant. If proven, a defendant may be fined $500,000 and receive up to 10 years in prison. Corporations found in violation may be fined as much as $5 million, and the government may seize any stolen secrets and property. The average time for trade secret litigation is 2–4 years, with an average cost of about $4.1 million (Trade secret litigation 101, thomsonreuters.com) (https://legaljobs.io/blog/patent-litigation-statistics/).

5 Biggest Challenges of Working in Biopharma IP

Biopharma companies are typically multifaceted, cross-functional organizations, so aligning scientists, engineers, and marketing individuals from different functional groups of the organization can be a challenge, particularly when there are time-sensitive deadlines to meet. Establishing a regular educational program that provides an overview of IP basics, the timing of patent filings prior to any public disclosure, and the need for appropriate IP-related agreements in place prior to working with others, is essential to keep tight control over IP.

6 Greatest Joys of Working in Biopharma IP

By far, the greatest joy is working with creative, bright, talented, and truly innovative scientists and engineers. Being involved in a career where each day there is something new to learn and being at the forefront of some of the most remarkable and revolutionary innovations that truly impact our daily lives is both an honor and a privilege.

7 Journey into Becoming a US Patent Attorney

Jas Sagoo emigrated with her family to the United Kingdom at the age of three due to Idi Amin's expulsion of Indian Asians from Uganda. Idi Amin was a Ugandan military officer and the president of Uganda. As an abusive dictator, he turned a harmonious country where Ugandans and Asians thrived together for many years into a country of political repression and ethnic persecution.

Her father, a teacher of science and mathematics in Kakira, saved his meager earnings as a schoolteacher and helped his parents, brother, and two sisters leave Uganda. Jas and her mother remained behind with her father for about a year before the environment got too dangerous for them to stay due to the rising violence of the Ugandan military. This was an extremely emotional and difficult time for the family. She recalls the day in December at Kakira airport when her mother was holding her hand to board the plane to the United Kingdom while she cried and struggled to run to her father, who was looking on from the platform above. As a child of three, it was a traumatic experience for her, leaving her father behind. He was contracted to remain to finish off the academic year in an increasingly uncertain environment, alone and away from all his family. He joined them in the United Kingdom as soon as he was able to.

Arriving at London Heathrow on a snowy December day, Jas and her mother had no warm clothes or winter shoes. She slipped coming down the stairs of the plane as they disembarked, but her mother was holding her hand tightly and stopped her from falling. They had never seen snow before nor felt the chill.

As immigrants in a new country, there was a big cultural difference, but the family persevered and overcame many challenges. The greatest issue was the language barrier. With a vocabulary of only "yes" and "no," she had to learn the English language ab initio before starting school (reception) at age four and a half. To encourage this, her parents insisted they speak English at home and try to integrate into the culture. Integration was a challenging, as many of the British people were not enamored by the influx of foreigners, even though Africa was part of the British Empire.

Jas was a naturally curious child and gravitated to learning whatever she could. She knew at an early age that she wanted to go to university and wanted a degree. When she was about seven, she asked her mother to make a tassel that she attached to the corner of a chocolate box to create a graduation cap. She used one of her mother's scarves tied around her neck as a graduation gown.

At the age of 16, she was always interested in both arts and sciences, inspired by her father. However, the United Kingdom school system at the time was designed to navigate students into either a science track or an arts track. She had selected A-level chemistry, biology, and English literature (A-levels being entry exams for university at age 18) but was quickly told by the school to drop English literature and select either physics or mathematics. She chose physics and thereafter was on the scientific path but still had an interest in law. To satisfy that interest, during the university holidays when she returned home, she volunteered to work with solicitors and barristers who provided free legal services at a Law Centre. She found this opportunity by simply visiting the Law Centre and offering her time. The very busy solicitors and barristers were very eager to have volunteers help them with background research on cases.

After completing her Bachelor of Science (B.S.) degree in biology and chemistry at Nottingham-Trent University, she was offered an opportunity to complete a Ph.D. program in biochemistry at the University of Nottingham. This further funneled her down a purely scientific track. As part of the Ph.D. program, she spent several years at a biotechnology company called Celltech (now Union Chimique Belge (UCB) pharmaceuticals). It was there that she learned about the role of an in-house patent agent, and the light bulb moment went off for her! The patent agent filed global patents protecting the innovations of Celltech and used his scientific background and legal arguments to get the patent granted in several different countries. This was what she had been looking for—a job that married both science and art.

After earning her Ph.D., she applied for several apprentice positions in the United Kingdom to learn under a qualified patent agent, a requirement necessary to take the European patent exam. She heard back from no one. At that time in the United Kingdom (and even today), her foreign name and the somewhat elitist profession she sought to enter were barriers to entry. This profession was and remains dominated primarily by British men with an Oxbridge education or connections in country clubs.

As she had no success entering the IP profession in the United Kingdom, she was advised by her Ph.D. supervisor to take a postdoctoral position that had been offered to her at Harvard Medical School/Dana Farber Cancer Institute. It was a difficult and emotional decision to make at age 24 to move more than 3,000 miles away from her family and everything she knew. Her parents encouraged her to take the opportunity and reassured her that she could return to the United Kingdom at any time. She was soon on a plane to Boston, feeling sad at leaving her family, anxious and nervous about the postdoctoral position, but also excited by the new adventure she was going to embark upon.

Now an immigrant in Massachusetts, she felt like she was starting from nothing again. She stayed with a relative who lived in Massachusetts for a few weeks and managed to save enough money for the first month's rent for an apartment of her own. The postdoctoral salary was paltry, and the environment at Harvard was competitive. Nevertheless, she managed to obtain a few publications from her research while still looking for opportunities to get into IP law.

Prior to the days of the internet, she spent hours at the Boston Public Library and made a list of all the law firms in Boston. She then called each one, asking to speak to anyone who could give her advice on how to get into patent law. Most of her calls were rejected by the busy law firm attorneys, but three law firms did speak with her. Two law firms called her in for interviews, and both offered her a job as a Technical Specialist. Persistence and perseverance did pay off in the end.

The Technical Specialists program started around the mid-1990s in several Boston law firms to support the influx of biotechnology companies entering the region. Many inventions needed someone who could understand the technology and translate it into a patent. At the law firm, she was trained to draft patent applications and respond to correspondence from the United States Patent and Trademark Office (USPTO), a federal government agency. After completing the patent bar exam, she was registered to practice with the USPTO. She also completed her Juris Doctor (J.D.) law degree after attending four years of evening classes while working in a busy law firm and finally qualified as a patent attorney. She was the first and only person in her family that earned a Ph.D. and J.D. Her dream was realized; she was finally a patent attorney working in IP law. However, this was merely the beginning.

Working in a busy private practice law firm in the heart of Boston exposed her to some remarkable scientists and inventors and taught her to work in several different areas of biotechnology, which were very different from what she had studied in her Ph.D. and postdoctoral research programs. She noticed, however, that to be completely entrenched in the IP strategy (which often follows a business strategy), she had to work closer with the scientists and inventors. This prompted her to take a role as a patent attorney in-house at a large Swiss pharmaceutical company, moving to Switzerland for two years to start her in-house career. In-house practice was very different from private practice. In private practice, attorneys are only exposed to discrete projects without visibility into the overall business strategy, whereas in-house, the attorney has visibility into the overall business strategy, provide relevant insights, and knows exactly how and where the discrete project fits into the bigger picture.

As an in-house patent attorney, she could truly help shape the IP strategy for a biopharma company and help the scientists navigate through complex IP situations. She is fortunate to have worked with scientists who revolutionized aspects of medicine. She has drafted patents for technologies such as Chimeric Antigen Receptor T cell (CAR T) therapy, where a cancer patient's own cells are engineered with a protein that seeks out and kills cancer cells in the patient, as well as unique instruments such as the Seahorse XFPro that measures cell metabolism of live cells.

Her journey took her across three continents and finally brought her to Agilent, where she now manages a team of patent practitioners. The moral of the story is never give up on your dream. Persevere, find a path forward, or make your own path.

8 Contracts Administration and Commercial

Lata Kirtivasan, Tsering Ringu, and Miranda Pierce

8.1 What Is Contract Administration and Commercial?

Contract Administration is the end-to-end management of contracts (the contract life cycle). The contract life cycle begins with understanding the goals of the enterprise business and its relationships with customers and partners. The Contracts Specialist receives a request (1) to create a new contract or (2) to amend an existing contract. The next step in the process depends on the company's guidelines for contract administration. The guidelines can include tasks such as performing due diligence, seeking clarification from Sales on the type of contract needed, and obtaining approvals. Once the Contracts Specialist understands the scope and aligns with the company guidelines, the contract request can be processed, and the contract documents are created and assembled as a draft. After drafting the contract, it is sent to the counterparty for evaluation, possible negotiation, and signature.

Contract Administration plays an integral role in a biopharma organization. As the overall gatekeeper of the company, they ensure the correct contract types are assigned as per the relationship with the customer/partner. After a contract is signed, the key responsibility is to administer the contract, send communication to all stakeholders, and set up discount or price management in the customer relationship management (CRM) tool.

The Contracts Specialist manages many different types of requests, including changes to terms (contract modifications or amendments) from partners, customers, or internal business stakeholders. Changes that have an impact on terms may include discounts, payment terms, or the length of the contract. These modifications require consulting with internal stakeholders and other departments within the organization, such as finance, sales, human resources, and others. When there is a request for changes to the Terms & Conditions of a contract, the contract administration team will work with the commercial legal team, who provide legal advice and solutions for any such changes. Consequently, there is close collaboration required with these functions, as well as a good understanding of organizational roles and responsibilities. Contracts Specialists perform semi-annual and annual reviews of the contracts they administer. Key responsibilities include analyzing the terms of the contract to understand if the contract obligations are met and communicating any changes with the business.

There are various roles within a contracts administration department needed to manage the end-to-end contract lifecycle, including business process analysts, project managers, and contracts specialists. Each role has a specific focus throughout the end-to-end contract management lifecycle. Detailed information about these roles can be found in the Sect. 8.4 below.

Commercial is a team of regionally focused lawyers who provide commercial legal advice and solutions to enable the business. This may include contract Terms & Conditions, advice on liability, management of competition law and advice, as well as support on sales and tender contracts. The team works collaboratively across the legal organization to ensure that the commercial organization is appropriately supported and works as the main point of contact for the legal needs of different business functions.

8.2 *What Do I Need to Study to Work in Contract Administration and Commercial?*

Any formal educational background within any area of study is acceptable to pursue a career in Contract Administration. To work in Contract Administration, a bachelor's degree or equivalent experience is usually required, and a master's degree is optional. A key driver in relation to education and a career in Contract Administration

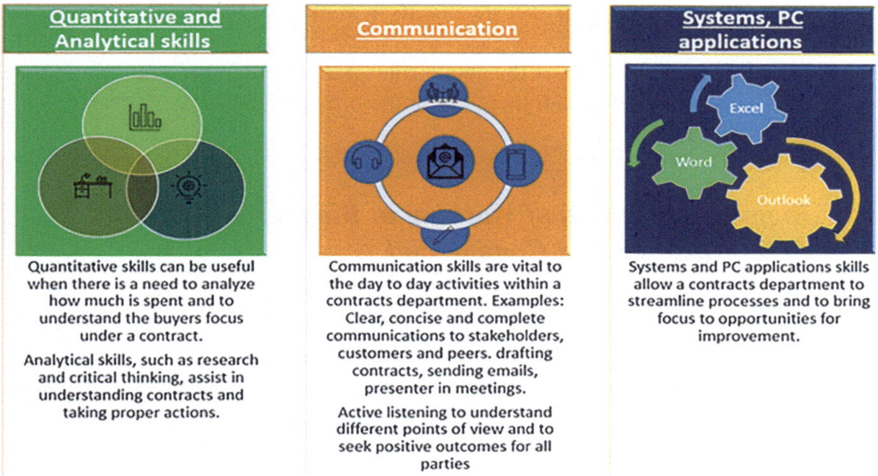

Fig. 5 Examples of contracts administration skills

is to keep your curiosity and continue to learn and develop new skills. You never know when that knowledge or skill will be an asset to your career.

To work in Commercial, a law degree is required for the region in which you want to practice law. Additionally, a background in business is preferable since the commercial team is primarily the point of contact for commercial/sales functions within an organization.

8.3 *What Skill Are Necessary to Perform in Contract Administration and Commercial?*

There are many skills utilized within a Contracts Administration and Commercial department. Fig. 5 highlights some of these skills. Each of these skills can be developed using existing resources found on the Internet.

8.4 *Different Specialties Within Contract Administration*

There are several types of roles in Contracts Administration, including:

- Contract Specialist: This role's focus is to manage contracts following the guidelines established by your company. Contract types managed by a Contracts Specialist can vary depending on the company's products or services and company's business model. For example, contracts can be local to a specific region or contracts can include various regions and be worldwide contracts.

- Business Process Analyst: This role's focus is to manage tools and processes for the Contracts team.
- Project Management: Both of the above roles participate in projects where they can lead projects or be a team member on the project.

8.5 Biggest Challenges of Working in Biopharma Contract Administration and Commercial

The biggest challenges of working in Biopharma Contract Administration and Commercial are the constant changes that occur during acquisitions or divestitures and changes to the portfolio when products are added/deleted/combined. These changes affect contracts, and they must be modified to align with such developments. When acquiring a company in a different industry, those working in Contract Administration and Commercial must learn about the new industry and use active listening with customers in the new industry to better understand the dynamics of that industry.

8.6 Greatest Joys of Working in Biopharma Contract Administration and Commercial

The only thing that is constant when working in Contract Administration and Commercial is continuous learning. The excitement in the job is forever since the process of acquiring new skills and knowledge over time helps one not only upskill but also it opens the door for new opportunities. Learning is not just limited to hard or technical skills but also soft skills such as collaboration, leadership, communication, and teamwork.

8.7 Our Journey in Working with Biopharma

Everyone has their own unique story of how they selected or ended up in their respective career path. Below are three journeys into the world of contract administration.

8.7.1 Journey 1

Lata Kirtivasan was always curious and was inclined toward mathematics and science since childhood. This keen interest led her to study physics as a major during her university years. As she pursued her STEM degree (Science, Technology,

Engineering, and Mathematics), she began working as a freelance copy editor at a publishing company. The books that she edited were around the topics of mathematics and science, which kept her engaged in this freelance work. As she completed her master's in physics from Miranda House, Delhi University, one of the top-ranked colleges in India, she was employed by a leading education-oriented company in their e-learning department. In this education-oriented company, where she joined as a team member, she not only learned to create online learning courses but also learned how these courses are an effective way to teach. Most of these courses were around science and technology, specifically computer science topics such as computer languages, software, coding, operating systems, and so on. Because of all she learned in this role, she decided to evaluate her knowledge and completed her Microsoft Certified Solutions Developer (MCSD) certification. The 10 years in this company taught her not only technology but also project management, people/team management, and vendor management.

Wanting to continue to learn and develop new skills, she joined Agilent as a Proposals Project Manager, where she worked extensively on presales activities specifically in the life sciences space, including biopharma. Her new job was interesting; she had to continue learning about the new company and its products—what they do, working with various departments, etc. The STEM degree helped her come up to speed quickly and soon she was working as a Contract Programs Manager. In this role, she managed standard templates for legal contracts and process transitions, and she led and managed multiple projects. Her mentor in the company really motivated and encouraged her to learn terms and conditions associated with contracts, as well as new skills such as how to draft a contract and negotiation skills. This was a learning for her and a new stepping-stone, where she began to understand Terms & Conditions. Moving ahead, she continued to receive more opportunities in the form of projects or roles. She now leads projects, including acting as a Contracts Mergers & Acquisitions (M&A) lead, manages contracts, and acts as a Contracts Data Privacy Champion. The Contracts M&A lead role requires her to understand the requirements of the acquired company and to effectively integrate contracts into the company while minimizing execution risk by using consistent but flexible integration processes and methodology. Similarly, she also learned about data privacy and its importance in her new role as the Contracts Data Privacy Champion.

The various roles have helped her develop the skills necessary to work with cross-regional and functional teams and really contributed toward her extensive learning. The opportunities in the Legal Contracts department at Agilent have enabled her to learn, practice discipline, structured thinking, and move on to take on more complex and challenging roles and/or projects. The learning comes in two forms for her—one is the hard skills, that is, the core competency required for that job or role, and the other is the soft skills that come mostly by interacting with various people such as project/functional teams, leaders, mentors (formal and informal). The skills she learned from the various people she interacted with through various projects and roles are some of her most valuable learnings. She has always looked at each of these interactions as a learning journey that has helped her to develop and sharpen skills on decision-making, problem-solving, situational awareness, communication, strategic thinking, leadership skills, collaboration, and so on. To conclude, continuous learning contributes toward self-development and enables her to be the best version of herself. Her advice to others, regardless of which career path they take, is to never stop learning.

8.7.2 Journey 2

Miranda Pierce's journey began shortly after applying to a staffing agency. The decision was made to work with a staffing agency because it was difficult to find a full-time permanent position under the economic conditions at the time. There was an open position at Agilent Technologies, Inc., a life sciences company, for a 12-month position as an administrative assistant that she accepted. While the administrative assistant position was not aligned with her undergraduate degree, a Bachelor of Science in Leadership and Consumer Economics from the University of Delaware, she learned many key skills and gained knowledge that helped to facilitate opportunities for future growth. Skills that were developed as an administrative assistant included communication, organizational skills, such as time management, setting goals, and prioritizing tasks. One of these skills was teamwork. Miranda was new to the company and was supporting a group of 25 individuals and a manager, learning to understand the team dynamics and the organizational goals and how best to contribute and support both. Within several months of starting the job, the company decided to move the administrative assistant from a temporary position to a permanent position, so she applied for the permanent position and was offered the job. As she continued to learn and gain knowledge in this position, more opportunities arose. She was given the opportunity to develop, maintain, and report department metrics to management. In addition, she was able to participate in high-level projects by developing career position plans based on job functions. One of her coworkers and informal mentors suggested that she should apply for one of the customer service jobs. She took the advice, applied, and started a new position in customer service. Her decision to take her career in this direction was to learn a different aspect of the company, to be customer-facing, and to engage with customers.

The customer service position brought new learnings for Miranda. In her new role, she was no longer just supporting internal employees; she was helping to support the company's customers by taking calls on the company's 800 line and placing customer orders for services. Communication skills were quickly developed through formal job training and on-the-job experience. Listening, empathy, and writing skills were the key focus of the formal job training program and are important for meeting customers' expectations and providing solutions. After a few years in customer service, she accepted an offer to be a regional team leader collaborating with the customer master data management team. In this position, she acquired decision-making skills by using quantitative and analytical data along with additional general system applications, such as Excel and Visio. It was also during this time as a regional team leader that one of her informal mentors reached out to her and suggested that she should apply for an open position in the contracts organization.

Mentors, both informal and formal, can be very beneficial. Over the course of Miranda's career journey, she had both informal and formal mentors to help guide and provide constructive feedback. Informal mentors can be senior-level colleagues, peers working in another department, or even someone working in a different industry. Formal mentors provide guidance and feedback in a structured setting and for specific topics. For instance, Miranda was able to schedule monthly meetings with her formal mentors to discuss specific legal topics, such as learning about indemnification and improving negotiation skills by shadowing on live negotiations with mentors.

Miranda continues her journey working in contracts administration. She is taking training on mentorship—learning skills so she can be an effective mentor to others. Her journey guidance to her younger self would be to stay inquisitive, keep gaining knowledge, and have many mentors.

8.7.3 Journey 3

Tsering Ringu's childhood was spent in the Himalayan region where everyone she knew was either a doctor, an engineer, a civil servant, a farmer, a politician, or a teacher. But deep down, even as a child, she knew that the world could not be restricted to these limited career options. In school, she was interested in science and represented her school in competitions within the district and state-level exhibitions and seminars where she won prizes. Throughout her school life, everyone, including her teachers, believed that she would either become a doctor or an engineer. She surprised everyone when she decided to pursue a degree in history. Everyone she knew gave unsolicited opinions about her life. Many said she was a failure because she was not choosing a conventional career path. The main reason behind the switch from science to history was because opinions and views were not limited in this choice of study, and she got admitted to Lady Shriram College for Women, one of the premier institutes of India. After getting a bachelor's degree in history, she had trouble deciding on whether to pursue further studies. As a stopgap, she started working in customer service for a leading Internet company. Even from an early age, she was a vivid communicator. Less than a year into working in customer service, she joined the website copywriting division. She enjoyed writing and consequently spent 4 years in that role. She then transitioned to a quality analyst role, which involved checking the copywriters' work. As she was transitioning to this role, she realized that it was not her field of interest. While looking for a new opportunity, her friend told her about a new position in proposal writing at Agilent. She joined Agilent in a presales function of contracts and was introduced to the world of STEM. Even though she did not have a STEM background, she seamlessly fit into the organization as she naturally had the skills to succeed in the job. The skills required were communication and writing. She was writing about life sciences, but the content for the technical portion of the proposal was provided by other departments. Since proposal deliveries are always on tight deadlines, she learned to negotiate with various departments in order to meet the deadline, something she excelled at. Because of these skills, she was moved to the Proposal Project Management team. In this new role, she negotiated with other departments within Agilent in order to get all of the information needed to finalize a proposal so that it was ready to submit to the customer. As she was working in this role, her informal mentor that worked in her group recognized her negotiation skills and encouraged her to use those skills professionally in the contracts world by giving her assignments involving Contracts Administration. She soon started supporting Contract Administration together with her responsibility of Proposal Project Management and, within a few years, left this role to become a full-time Contracts Specialist. After 11 years in this role, there is no lack of new challenges because of the complexities of the job. There are different kinds of requirements from various stakeholders, which modify the Contract documents to reflect these changes and enable the Contracts Specialist to think in a creative way to implement these changes.

Her journey so far has made her realize that if you stick with your gut, persevere, and ignore the outside noise, eventually, you end up doing something you enjoy in a field that you and the people around you never knew existed.

9 Legal Compliance, Privacy, and Employment Law

Virginia MacSuibhne, Leslie Stevens, and Shannon Jamison

9.1 What Is Compliance?

Compliance programs are the newest addition to biopharma organizations and may be found inside or outside of legal departments. Corporate compliance programs have their roots in the 1991 United States Sentencing Commission issuing of the US Federal Sentencing Guidelines for Organizational Crime, which established that an effective compliance program could lessen criminal penalties for an organization. The 2002 passage of Sarbanes-Oxley, and some significant corporate accounting scandals like WorldCom, Enron, Tyco, Lehman Brothers, the Madoff Ponzi scheme, and Freddie Mac, as well as increased enforcement of the Foreign Corrupt Practices Act (FCPA) [1] in the early 2000s, made corporations to adopt programs developed around the US Department of Justice's 7 Elements of an Effective Compliance

Fig. 6 Seven elements of an effective compliance program

Program framework (Fig. 6). Today, such programs are found in every publicly traded company and many privately held organizations.

9.2 *What Do I Need to Study to Work in Compliance?*

As compliance is still a relatively new field (less than 25 years old outside of the financial industry), the backgrounds of those in it vary greatly; some are lawyers, some come from regulatory or quality, some start in audit or fraud examination, and others come from business backgrounds such as operations, marketing, or sales, or have special skills in project management, adult education, or communications. There is no one "right" path into compliance. There is also no STEM background required, though interest in and the ability to learn and understand the basics of how different technologies and products work is always appreciated by the business and valuable for any role in a biopharma, life sciences, or healthcare technology organization.

Now, more and more, there are various educational programs being offered specific to compliance, including several master's programs. Most compliance professionals get a certificate as a Certified Compliance and Ethics Specialist (and maintain continuing education) with an organization such as the Society of Corporate Compliance and Ethics/Healthcare Compliance Association (SCCE/HCCA) or, in the case of privacy-focused professionals, the International Association of Privacy Professionals (IAPP).

9.3 Different Specialties and Key Topics Within Compliance

Compliance Departments are responsible for advising on a variety of different areas of risk including some general areas like compliance operations, third-party risk management, and key topics like anti-bribery and anti-corruption (ABAC), competition law, healthcare, regulatory, trade compliance, privacy and cybersecurity, employment claims, and litigation. Generally, compliance programs are designed to evaluate risks based on the location, activities, and other aspects of the business and then mitigate those risks. Compliance programs are focused on preventing and detecting wrongdoing.

9.4 Compliance: Operations, Third-Party Risk Management, Key Topics (Anti-bribery and Anti-corruption, Competition Law, Healthcare, Regulatory, Trade Compliance)

9.4.1 Compliance Operations (Policies, Helpline, Communications, Training)

Compliance Operations is responsible for building the infrastructure and maintenance of an effective compliance program. To meet the requirements of regulators and industry standards, an effective compliance program must include policies and procedures that are written and communicated to employees about appropriate and legal behavior and business practices. These policies can be anything ranging from a standards of business conduct that includes brief descriptions of key legal and compliance topics and sets the tone for employees on legal and ethical business practices, to specific policies on topics such as conflicts of interest or anti-bribery and anti-corruption. Compliance Operations is then responsible for training on those policies to ensure understanding and providing a mechanism for reporting policy violations and misconduct (a compliance helpline, for example).

The work of a Compliance Operations professional does not stop at the infrastructure of policies and training; it also includes ongoing communications to the business and support on key compliance issues. For example, every year in the fall is Compliance Week, a week dedicated to promoting compliance globally by engaging with employees through innovative communications and activities. This requirement for continued communication and connection to the business is key to an effective compliance program.

9.4.2 Third-Party Management

One of the major risks of any compliance program is doing business via third parties such as distributors, resellers, or other intermediaries. Therefore, a large component of any compliance program is understanding the third parties an organization does business with and managing and mitigating the risks by conducting due diligence, background reviews and investigations, ensuring contract terms and conditions clearly state expectations and consequences for noncompliance, and engaging in auditing, monitoring, and compliance checks to ensure compliance to laws and regulations by third parties. The "watch-out" areas for third parties are bribery and corruption, competition law, regulatory, and trade compliance.

9.4.3 Key Compliance Topics: A Quick Overview

Some of the key topic areas a biopharma compliance team will be responsible for covering and understanding include anti-bribery and anti-corruption, competition/antitrust law, healthcare laws and regulations, and trade compliance. Anti-bribery and anti-corruption (ABAC) laws and regulations include the United States Foreign Corrupt Practices Act (FCPA) 15 U.S.C. Sec. 78dd-1 et. Seq., the Criminal Law of the People's Republic of China (the PRC Criminal Law), the UK Bribery Act [2], the OECD Anti-Bribery Convention [3], and similar laws making bribery or corruption of foreign government officials and others punishable under criminal and civil statutes. Competition and antitrust laws are designed to promote free and fair competition in markets by regulating anticompetitive behaviors like predatory pricing, monopolies, tying, price gouging, refusal to deal, and cartels. Healthcare compliance covers topics ranging from regulatory requirements in the healthcare industry such as Good Manufacturing Practices (GMP) [4], Good Clinical Practices (GCP) [5], and Good Laboratory Practices (GLP) [6] to International Standards Organization requirements, regulatory body requirements such as the United States Food and Drug Administration (FDA), the European Medicines Agency (EMA) in the European Union, the National Medical Products Administration (NMPA) in China, and similar agencies throughout the world. In addition, trade compliance will also be an important consideration for most biopharma companies, including product classification under export controls and screening of both customers and suppliers to ensure that things are not being supplied to persons, companies, or countries on sanctioned parties lists.

Fig. 7 Compliance programs

Most compliance professionals are familiar with some, but not all, of these regulations. Some choose to become specialists in areas such as healthcare and regulatory compliance, while others are generalists. Generalists focus more on the basic principles of ABAC, competition law, and the framework of a compliance program, including fundamental components such as investigations, policy, or matters that may require more expertise (Fig. 7).

9.5 *Key Activities*

9.6 *Privacy and Cybersecurity*

Barely a month goes by these days that there is not a new privacy or cybersecurity law being considered or passed somewhere around the world. Our job is to be informed and understand the potential impact on our business. This includes understanding the impact on how our company wants to and needs to transfer data and human materials around the world for research, development, and business purposes. It also includes developing programs to ensure that we can meet the obligations the law requires in terms of things like data subject rights, breach notification, security to enable privacy, and privacy by design. Given the proliferation of privacy and cybersecurity laws around the world, privacy and compliance professionals must simplify for their companies the myriad of requirements into key concepts and principles. Privacy and cybersecurity overlap when it comes to protecting *personal*

Fig. 8 Protecting personal data

data. Typically, privacy is managed within legal departments, and that team provides legal guidance to Information Technology (IT) Security functions when it comes to topics like security incidents, data breach notification, cyber insurance, negotiating IT Security agreements, etc.

Personal data is *one type* of sensitive data that requires appropriate protection through a cross-functionally designed governance program, but often privacy programs will have, within their remit, the protection of other types of sensitive information, including more broadly, confidential information (Fig. 8).

Privacy professionals at biopharma companies are tasked with distilling down those laws that impact where and how their companies sell products and services around the world. Like compliance, privacy must be managed at a programmatic level with policies, procedures, risk assessments, and monitoring. A large part of the job in recent years is building a program where there used to be none (legal departments generally have not operated as "programs"), as well as tracking constant updates to the regulatory landscape. At its core, most privacy and cybersecurity laws require that companies do three things: be transparent, be fair, and design cybersecurity and privacy principles into the way work is done (through clear, easy, and actionable guidelines, processes, and procedures). Successful privacy professionals will understand that the common principles of transparency, fairness, and privacy and security by design should be embedded into how their biopharma companies operate. By focusing on key principles, privacy professionals will be able to create scalable (and understandable) programs that can combat most, if not all, new privacy and cybersecurity laws coming down the endless pipeline of new regulations.

Like compliance, privacy is a relatively new specialty for in-house biopharma legal teams, at least when it comes to the need to establish a risk-based, documented program. Many privacy professionals come from legal backgrounds, having learned privacy "on the job" while negotiating commercial agreements with an increasing focus and negotiation time spent on privacy clauses. Others come to privacy from adjacent specialisms such as IT Security or simply from having worked on companywide privacy regulation implementation projects. Privacy and cybersecurity are about understanding what data a business collects, processes, stores, and uses, and how to do so transparently, fairly, and securely. Therefore, curiosity and an enthusiasm for building meaningful relationships across all functions is what is truly required to be a successful in-house privacy professional.

9.7 Employment Law and Employment Claims and Litigation

Every organization with employees must consider the employment laws in the jurisdictions it operates in and ensure compliance with those laws and regulations, which range from compensation and benefits (e.g., minimum wage, required breaks, vacation, overtime, and holiday pay) to policies and procedures (e.g., required postings, safety, discrimination, harassment, bullying, and managing complaints). Countries, states, provinces, and even counties may have regional or local differences in these important employment matters, and employers must understand and comply with the different regulations where they operate.

There are many varieties of employment frameworks, including at-will employment, contractual relationships, unions, or collective bargaining arrangements, and works councils. Additionally, there are different agencies or courts where current or former employees may file claims or litigation when they feel their rights in these areas have been violated by employers. Therefore, most companies have an in-house team dedicated to compliance with different employment laws, regulations, and filings, as well as managing disputes, claims, and litigation by current or former employees.

Having policies (such as those against harassment, discrimination, bullying, and retaliation), training for employees and managers, and procedures for reporting concerns or complaints of possible misconduct, and ensuring those concerns are properly investigated, can be used as a corporate defense in certain claims by employees of misconduct. Employment law teams often support these different policies, training programs, and procedures, as well as the defense of claims.

9.8 Biggest Challenges of Working in Biopharma Compliance

One of the biggest challenges for biopharma legal teams, particularly for compliance and privacy, is the constantly shifting regulatory landscape. Not a single day passes without a new law being passed, considered, or amended somewhere around the globe, and many of these laws are different from each other and/or do not yet reflect the state of technology, leaving grey areas and room for interpretation. This is particularly the case where we see the law trying to catch up with technological developments often adopted by biopharma. One current example of this is the burgeoning draft regulations on Artificial Intelligence.

The constantly shifting regulatory landscape means that in-house biopharma legal teams must constantly monitor and assess these regulatory developments across various jurisdictions and adapt their programs accordingly. This is particularly challenging for biopharma given that most legislation is drafted with a business-to-consumer company focus in mind (such as social media and big technology companies), not business-to-business organizations. As an example, little practical guidance is available for how to implement burgeoning privacy legislation in the biopharma and business-to-business context. This makes networking, being active participants in professional association groups, and conducting informal benchmarking critical to staying on top of this changeable area of law.

9.9 Journeys in Compliance

9.9.1 Journey 1

Virginia MacSuibhne started her path to compliance as an undergraduate designing her own degree in social justice with a combination of courses in sociology, anthropology, women's studies, criminal justice, and communications. She began a part-time job as a court runner for a local law firm, where she learned the ins and outs of the work life of attorneys in a small local practice in San Jose, California. She applied for law school and was the first and only attorney in her family. During her second summer, she participated in on-campus interviews and landed a summer internship with a larger law firm in the Silicon Valley of California. She was offered a job to join the firm after graduation and watched the office grow from 30 to over 300 attorneys during her time with the firm. She worked on a variety of different matters during her summer internship, from licensing to research and writing to document review and motion and trial preparation. Upon joining the firm, her primary focus was on litigation and disputes, as opposed to transactions and deals. During her time with the firm, she reviewed documents, wrote memos and motions, researched, took depositions of witnesses, did several arbitrations, wrote a brief for the Ninth Circuit Court of Appeals, appeared in court for hearings and trials, reviewed more documents, and even got to write a brief for the United States Supreme Court.

Much of her work ended up being focused on defending employment matters such as wage claims, discrimination, harassment, and/or retaliation claims, and defending securities fraud cases. Both subjects formed the basis for what is now the compliance framework. Specifically, employment law defense for an organization is founded on the idea that when organizations have robust mechanisms for reporting concerns, strong policies and procedures preventing improper conduct, and effective training and communications that teach employees and managers expectations preventing improper conduct in these areas, these factors will be considered in defense of the allegations. Likewise, securities laws such as Sarbanes-Oxley provide a similar framework for creating an effective organizational defense for misconduct; if there are reporting mechanisms, strong policies and procedures, and effective training and communications, these factors can mitigate an organization's liability when misconduct is detected. Both of these experiences were influential to her in terms of why and how compliance programs, and the seven elements that later were identified by the Department of Justice, resonated as meaningful and familiar.

After 6 years with the firm, she went to a publicly traded software company and focused on supporting their employment law needs. She worked closely with the human resources team, counseling on a variety of employment issues including compensation, benefits, leaves of absence, accommodations, policies, investigations, trade secrets, preventing harassment, discrimination, retaliation, and workplace violence prevention and safety. She provided training to employees and managers, helped structure programs, policies, and procedures to mitigate the risk of employee disputes, and to manage and resolve employee claims and litigation. During this time, she was also asked to work on the company's anti-bribery policy and the first code of conduct for the organization, further evolving her work in compliance and seeing where organizations were beginning to consider developing compliance programs. At the time, not all companies had yet adopted the idea that the role of Compliance Officer was a necessary and separate one from the General Counsel.

Following this, she took those experiences and moved into the biotech sector, becoming the first compliance attorney for a local bio company. When interviewed, others on the team asked if she thought compliance was really a full-time role. Thankfully, her manager saw the value and the future of compliance. Virginia's new role not only was a full-time role but eventually, as laws like the Sunshine Transparency Law and different privacy laws passed around the world, a growing industry and department where she was able to become a manager, then a leader, and an executive.

She spent 2 years as the General Counsel and Head of Legal for part of the company in a different state, had the opportunity to travel to many parts of the world, did a six-month visit to Switzerland on assignment, led various projects (unrelated to legal or compliance), and had many leadership training courses because of her work with the organization. In addition to her professional life, she used nonprofit volunteer work as a low-stakes way to build leadership and communication skills outside of the work environment.

Willingness to continue to learn, raise your hand for opportunities, and try new things using skills like communication, project management, meeting facilitation, storytelling, presenting, organizing large amounts of data into concise pieces, and engaging a team are all more valuable than subject matter expertise in an organization. Once you are in the biotech space, networking with others in the industry through industry groups is a great way to connect, benchmark with peers, and hear about potential future job openings. Often, breaking into the field will be the hardest part, but once you have had a role, if you stay current on the laws and regulations and, in the case of compliance, keep up with your compliance and/or legal continuing education, you should be well-positioned to find other future roles in the same or adjacent spaces.

9.9.2 Journey 2

Shannon Jamison's path to a career in compliance was not a traditional one. From when she was very young, Shannon always excelled in writing and storytelling. She attended California Polytechnic State University, San Luis Obispo, and got her degree in Journalism with the intent to work in Public Relations (PR). She had always enjoyed storytelling, and PR was a natural fit for her skills and creative personality. In her first role at a PR agency, Shannon worked with clients in the technology sector, specifically semiconductor companies. She quickly had to learn and understand complicated technology and translate it for a lay audience, a skill she would take with her to every job she has had since. In fact, when Shannon got her first job in biotech, it was one of the reasons she was hired—because she could take complex information and make it easy to understand.

On Shannon's path to compliance, she faced a few hurdles, including a manager who told her that her career growth would slow down because she had just had her first child. The manager's exact words were, "You are on the mommy track now. You won't get the same opportunities as you did before." But Shannon refused to let that stop her. If anything, it ignited a fire in Shannon to prove that manager wrong, but also raise concerns and use her voice when she experienced or saw inequalities in the workplace. While she did not know it then, she was already practicing compliance while not being in a compliance role.

After working in PR agencies for about 7 years, Shannon was ready for a change and took a role as an Internal Communications Manager for a semiconductor company. It was there that Shannon started to feel disconnected from her work in technology and was craving that feeling of connection and purpose in her work. She saw a role posted at a biotech company for a marketing communications manager, and she leapt at the chance. That role, as the marketing communications manager for a portfolio of genomics and oncology diagnostics, changed Shannon's life and career path forever.

When Shannon took on the challenge of working in biotech, a new industry sector, she immediately realized what she had been missing. Working in biotech, promoting products and services that help people, was the connection to a purpose that Shannon had been missing in her career. She got to interact with patients who were directly impacted by the company's products and have conversations with some of the brightest healthcare professionals and key opinion leaders about what would be the next advancement in science to help people.

It was also during her first role in biotech that Shannon even heard the word "Compliance" mentioned. When Shannon was in PR or working at semiconductor companies, compliance wasn't a role. So, when she was still new in her position in biotech and the Compliance Officer came around asking for some documentation, Shannon didn't really know what to do. But what Shannon did learn that day was how important compliance was and still is in biotech.

It wasn't long after that Shannon had more interactions with the Compliance Officer. Following a minor infraction from some of her team members, Shannon was asked to help her manager with a compliance project with the Compliance Officer. They were tasked with working on training and education for a new policy on interactions with healthcare professionals. The project went incredibly well, and it was then that Shannon was approached by the Compliance Officer to consider a career in compliance. The Compliance Officer recognized the importance of Shannon's skills in storytelling and her ability to take complex information and translate it for a broader audience. It was exactly what the function of compliance needed, and she said Shannon should consider a career change.

Fast forward, Shannon not only considered a career change at that moment, but compliance became something she realized she also had a passion for. Working in compliance is so much more than just laws and regulations; it is about doing what is right for the safety of the population. It is about good business practices and integrity—things that aligned with Shannon's core values. When the job became available to be a Compliance Training and Communications Manager, Shannon took the leap and never looked back.

Now, as an Associate Vice President of Compliance Operations, Shannon feels at home in her career. She found a purpose working in compliance. In addition to on-the-job training, Shannon also decided to pursue a master's in legal studies for compliance and ethics, solidifying her career path and unlocking new opportunities.

9.9.3 Journey 3

Leslie Stevens began her path into compliance, specifically data protection and privacy law, while working for the Irish Government's Investment and Develop Agency (IDA Ireland) during the day and attending law school (Santa Clara University School of Law) at night. During her time working with IDA Ireland, she heard first-hand the questions and concerns the burgeoning technology sector in Silicon Valley had when considering placement of operations in a country and region which would subject them to stringent data protection and privacy laws (albeit these were the days of the Data Protection Directive 95/46/EC [7] as opposed to the General Data Protection Regulation 2016/79) [8]. At this time, with the rise of social media companies in Silicon Valley and their increasing (and unregulated) collection of personal data in the United States, Leslie became curious to understand exactly how these companies could operate their personal data-driven business models within the European Union, where privacy was recognized as a human right. Combining the interests piqued from her day job with her evening law degree, Leslie pursued a High-Tech Law Certificate and focused her thesis on privacy. Because Leslie worked a full-time job during the day, she missed out on the traditional summer internships with law firms and knew she had to get creative with her next steps. Like

Virginia's journey to Chief Compliance Officer, Leslie would also be the first attorney in her family (and broader social circle). This meant she had little understanding of what a legal career path would look like for her but was determined to find one.

So, it was upon graduation and after taking the California Bar Exam that Leslie applied to and was accepted into a Master of Laws (LLM) degree program at the University of Edinburgh, Scotland. She knew she wanted to understand all there was to know about data protection and privacy law and figured that being in Europe (the birthplace of privacy regulation) was the best place to do this! While pursuing her LLM, Leslie ensured that she networked with her supervisors and other staff (despite her true introverted self). She did this to make her enthusiasm for data protection and privacy known (and hopefully to be included when any opportunities for work arose). During this time, Leslie was approached by her LLM supervisor and future mentor to work on a consultancy project understanding why researchers in the United Kingdom found it difficult to access data for their work, particularly when their research was pursuing public interest topics such as enhancing health outcomes for the community. This first (paid!) opportunity led Leslie to more opportunities, including a scholarship to complete a PhD in Data Protection and Privacy Law and a job as a Research Fellow, where she continued her work on understanding the intersection of public interest research and the regulatory framework of data protection and privacy law—seeking to demonstrate privacy law as an enabler for such research. Leslie completed her PhD and Research Fellowship in Scotland, only to find, by chance, that a company from her home in California (Agilent) was hiring a privacy attorney based in the United Kingdom to help them comply with the newest law on the block—the EU General Data Protection Regulation (GDPR).

The rest is history—6 years later, and at the time of writing this chapter, Leslie is still at Agilent (but back in California!) and now Associate Vice President, Chief Privacy Officer within the Legal Compliance Function. This is a role Leslie could only dream of back when she was pursuing her Juris Doctor (JD) and wondering if practicing law in this (or any) topic she loved would ever be possible. Despite Leslie's atypical legal background (no summer internships, no fancy law firm experience), her curiosity, true passion for data protection and privacy, and tenacity to follow through and not give up despite the obstacles (or oceans) in the way led her to her current role, which to this day continues to excite and challenge her. Whatever limitations Leslie may or may not have had, she did not let those define her, and that is at least one important reason she is where she is today—happily working in a fantastic biopharma compliance department.

9.10 Greatest Joys of Working in BioPharma Compliance

Working in a biopharma legal compliance department means working for a mission-driven company that can provide inspiration and fulfillment for the long term. It means that the work in-house teams contribute to daily supports the development of products and services that truly improve the quality of individuals' lives. The programs, procedures, and policies biopharma in-house teams establish ensure that products and services are delivered in a way that is ethical, responsible, and in line with the company's values. The work of an in-house biopharma legal professional supports the integrity of *how* a company does business. When that company is focused on advancing the quality of life, this makes the role infinitely easier, more enjoyable, and fulfilling.

References

1. Foreign Corrupt Practices Act: United States. (1977). *Foreign Corrupt Practices Act of 1977, as amended, 15 U.S.C. §§ 78dd-1, et seq. 11.*
2. UK Anti Bribery Act: United Kingdom. (2010). *Bribery Act 2010.*
3. The OECD Anti-Bribery Convention: OECD. (1997). *Convention on combating bribery of foreign public officials in international business transactions.*
4. Good Manufacturing Practices: U.S. Food and Drug Administration. (2016). *21 CFR Part 110 – Current good manufacturing practice in manufacturing, packing, or holding human food.*
5. Good Clinical Practices: Sarzotti-Kelsoe, M., Cox, J., Cleland, N., Denny, T., Hural, J., Needham, L., et al. (2009). *Evaluation and recommendations on good clinical laboratory practice guidelines for phase I–III clinical trials.*
6. Good Laboratory Practices: U.S. Food and Drug Administration. (1978). *21 CFR Part 58 – Good laboratory practice for nonclinical laboratory studies.*
7. Data Protection Directive 95/46/EC: European Parliament and Council of the European Union. (1995). *Directive 95/46/EC of the European Parliament and of the Council of 24 October 1995 on the protection of individuals with regard to the processing of personal data and on the free movement of such data.*
8. GDPR 2016/679: European Parliament and Council of the European Union. (2016). *Regulation (EU) 2016/679 of the European Parliament and of the Council of 27 April 2016 on the protection of natural persons with regard to the processing of personal data and on the free movement of such data, and repealing Directive 95/46/EC (General Data Protection Regulation) (Text with EEA relevance).*

The Sales and Management Professions

Risks and Rewards in the Industry

Rozaleen Dash and Ankita Bhargava

Contents

1 Introduction

The biopharmaceutical sector across the world is witnessing a robust growth trajectory and has proven to be highly inventive. Today, the global biopharma market stands at $383.23 billion and is expected to reach $554.03 billion by 2028, increasing at a Compound Annual Growth Rate (CAGR) of 7.65% [1]. The rise in

R. Dash
DBT Center of Excellence for Biopharmaceutical Technology, Indian Institute of Technology Delhi, Delhi, India

A. Bhargava (✉)
Business Development, Gennova Biopharmaceuticals Ltd., Pune, India
e-mail: ankita.bhargava@gennova.bio

S. Taylor, M. Campbell (eds.), *Women in Biopharma*, Women in Engineering and Science, https://doi.org/10.1007/978-3-031-80348-2_4

strategic collaborations (international and domestic) and investments in the biopharma world is only going to accelerate this growth. The huge demand for biopharmaceuticals is driven by many factors, such as an increase in the geriatric population and a rise in the prevalence of chronic diseases. Vulnerable diseases like cancer cause the highest number of deaths worldwide annually. As a result, there is a massive demand for biopharmaceuticals that have the potential to cure the disease entirely, hence increasing their acceptance worldwide.

Biopharmaceuticals have shown revolutionary advancements in the healthcare sector. The word itself consists of "biotechnology" and "pharmaceutical," two different fields that have combined to generate new opportunities for the biopharma industry. The primary difference between biopharmaceuticals and pharmaceuticals is the method by which the drugs are manufactured. The former is produced in living organisms such as bacteria, yeast, and mammalian cells, whereas the latter is produced through a series of chemical syntheses [2].

Since their inception, both biopharma and biotech companies have come a long way and continue to develop more reliable and efficient biomedicines for the welfare of society. Also, biopharma firms are equally trying to depose the paradigm of healthcare with their top-notch innovations and products. The biopharma landscape is rapidly growing, and a new fount of innovations is offering patients hope with each new development. The year 2021 has brought tremendous progress to the biopharma industry [3]. Novel platforms such as mRNA technology have been able to innovate and develop mRNA vaccines for the COVID-19 disease. As a result, the mRNA platform has shown the criticality of national and international collaboration within the biopharma ecosystem, and the necessity for a strong association between suppliers and manufacturers as continuous support for success now and in the future. As the biopharma industry continues to evolve post-pandemic, companies are now focusing on resilience, innovation, and long-term growth strategies to navigate emerging challenges and opportunities [4].

1.1 Biggest Challenges to the Industry and How to Maximize Productivity

Initially, the biopharma companies were focused on developing products for infectious diseases. However, in the past few decades, the focus has shifted to developing products for treating lifestyle diseases and chronic ailments. This has resulted in better tolerance of therapies among patients and much better outcomes over the years. Biotherapeutic drugs are among the most exquisite achievements of modern science. The complex structures of these drugs do not just look extraordinary in 3D modeling systems but also offer high efficacy and fewer side effects [5]. The efficacy and safety of biopharmaceutical products, combined with their ability to address previously untreatable conditions, allow pharma companies to command high prices for innovative drugs. There is much more to come in this field as existing treatment prototypes continue to evolve and become more sophisticated and as research is redesigned into entirely new types of products.

Over the past couple of years, there have been many challenges for biopharma companies. The most significant challenge has been the global pandemic, which created an unprecedented demand for vaccines, resulting in the cessation of many clinical trials, a transition from onsite to hybrid clinical trials, and disruptions in the supply chain. The industry has adapted and overcome many of these hurdles, and overall, business is thriving. At the time of this writing, June of 2023, the industry continues to drive innovative solutions while overcoming a backlog of orders [6–8]. Despite this success, biopharmaceutical companies face additional challenges, including meeting the demands of increasingly cost-constrained healthcare systems, major loss of revenue owing to patent expiry, and more demanding regulatory requirements.

In our point of view, the key challenge to overcome such issues for the biopharmaceutical industry is to substantially increase the number and quality of new, innovative, and cost-effective medicines without incurring R&D costs. However, it is widely recognized that the latest trends in biopharma R&D productivity have been moving in different directions for several years. As of the pre-pandemic era, the global biopharmaceutical industry was spending approximately $150 billion every year on research and development, but sales and profits continue to decline [9]. Many manufacturing companies have adopted technologies to innovate their processes and increase sales and productivity, but many others are behind in the adoption of these technologies. In the current scenario, the biopharma industry is moving from the scientific frontier to the business mainstream and will increasingly be forced to confront the same challenges faced by other businesses in terms of maintaining competitiveness by ensuring affordability, quality, and delivery performance. Let's discuss the most significant challenges the biopharmaceutical manufacturing industry has been facing and where there are growth opportunities.

1.2 Challenges and Opportunities in the Biopharma Industry

The biopharmaceutical industry is essential in terms of both employment and economic balance. However, it carries many challenges in addressing the development and manufacturing of advanced therapies. This section describes the challenges faced by the strategic sector, how to overcome them, and newer opportunities for the same.

Like nearly every other industry, the biopharma industry faces supply chain problems that witnessed unprecedented disruption all around the world [10]. These issues were worsened by the COVID-19 pandemic. Manufacturing production delays can be devastating—even one plant can have ripple effects across an entire supply chain. Therefore, to reduce challenges in the supply chain, companies need to implement changes to their strategy in order to maximize resources.

Industry 4.0 is called the latest industrial revolution and is expected to transform businesses using automation and data analytics to increase efficiency, reduce errors, and improve yield. Now, these principles are being adopted in the bioprocessing sector to become "Biopharma 4.0." Bioprocessing has relied on batch production for years, but many in the industry would like to see a move toward continuous processing to increase productivity with less drift. Regulatory bodies are in favor of increasing the use of automation and Biopharma 4.0 technologies. There is a demand for more investment in monitoring software to improve efficiency, even though we are in the midst of a global digital era. An end-to-end connected bioprocess is required, which is a key component of Biopharma 4.0, for real-time analytics, process efficiency, and predictability [10, 11]. All systems need to be connected digitally to run, control, and improve the process via machine learning/artificial intelligence. Our major challenge is to implement technologies now to keep up with demand and prepare as US FDA guidelines turn into regulations. Current guidelines recommend integrating quality systems and risk management approaches into existing programs to encourage the industry to adopt modern and innovative manufacturing technologies.

The biopharmaceutical industry needs a qualified workforce that has significant knowledge, experience, and skills. A trained workforce has the skills necessary to ensure, enhance, and improve their participation in their daily tasks, which will also help to fill in the skills gaps. To overcome this: (1) broad availability of good-quality education for future training and a close matching of skills supply to the needs of enterprises and labor markets, (2) enabling workers and enterprises to adjust to changes in technology and markets, and (3) anticipating and preparing for the skills are required. This is one of the essential factors that play a substantial role in improving productivity and ensuring a promising future in the biopharmaceutical industry [12]. Commercial strategies aim to improve the efficiency of employees and operations by optimizing the workforce. It should be noted that in the biopharmaceutical industry, optimizing the workforce maximizes employee benefits in terms of career growth and work satisfaction and thus maintains efficiency. In terms of batch production through batch folders, the biopharmaceutical industry is sometimes obliged to implement new standard operating procedures in operations and quality. For other areas related to the biopharmaceutical sector, there is a growing need for faster, more accurate, and transparent operations. In fact, it is essential to build a cohesive and communicative team rather than having a team of experts working in isolation. It is also important to involve the customer in biopharma product development. Moreover, initial planning and the product structure must be flexible to allow the evolution of customer demand throughout biopharmaceutical manufacturing [12, 13]. Finally, it is important that the biopharma product meets the implemented standards. Increasing the level of transparency has always been a topic of debate in the research and development sector. In addition, there was clear public support and interest in the biopharma industry during the pandemic, and it was necessary to communicate this openly to the world. The most obvious change is related to vaccines in the COVID-19 pandemic. Much effort has been put into keeping the patient's privacy while sharing information for clinical trials. It clears the way for great advances in science and medicine.

Consequently, there needs to be a systematic way to collect and review information so the biopharma product development process can be more productive. The Covid-19 pandemic resulted in a worldwide challenge without an evident solution. Many decision-makers and authorities relied on available data and established expertise to make decisions under time pressure. In addition to this, drug availability is not only the responsibility of the biopharma industry, since decision-makers are the ones who have the final say on whether or not a specific medication enters the market of a country.

1.3 How to Maximize Productivity in the Biopharma Industry

Biopharmaceutical industries are subject to strict standards and regulations and must carefully protect the integrity of their data. Therefore, they need to find a reliable and competent way to boost productivity and quality while maximizing productivity. To harness the full potential of new processes and become digitized, companies should communicate their strategy throughout the organization, from the labs to the factory to the executive level [14].

Switching to electronic data recording can increase efficiency and reliability in quality control. Initial results become available during the production process, and all the documents, including reports and compliances (as the act of conforming and yielding, in correlation with written rules and regulations or spoken), are ready as soon as a batch is finished. Electronic, automatic data recording also reduces the likelihood of errors when recording production data, compared to paper-based data recording [15].

In the biopharmaceutical industry, one small mistake can ruin a whole batch, and that leads to a significant amount of wasted money, time, and raw materials. If any unsatisfactory batch is missed by the manufacturer, the damage could be even worse—potentially leading to serious injuries and damaging the firm's reputation. Effective batch control is essential to avoid these negative scenarios. The data that manufacturers collect is crucial to ensuring product quality. The use of automation devices, such as robotics to automate multiple processes in drug development, helps to make data collection more efficient and reduce the likelihood of these errors. Efficiency refers to the ratio of input to output, and it is very closely linked to productivity, which measures the output as a function of total inputs [14, 15]. Validation ensures product consistency and safety and is a crucial part of the biopharmaceutical manufacturing process because of the regulations and standards the industry must adhere to.

Biopharmaceutical manufacturing processes are complex, with a wide range of equipment, systems, and facilities. A tightly controlled environment is required for the production process in biopharmaceuticals. Integrating advanced systems can make it easier to control the production process and environment, ensuring its optimization to increase productivity.

Improving the life and uptime of machinery in the biopharmaceutical manufacturing unit can increase efficiency. Investing in high-quality equipment can help industries avoid issues related to product development. By investing in high-quality equipment and creating an efficient maintenance program, businesses can avoid downtime and errors in production, thereby increasing productivity [15].

Security threats and cyberattacks can bring production to a halt. For example, phishing is one of the most prevalent cybersecurity threats, accounting for 85% of the threats to biopharma manufacturing processes. This is one of the reasons biopharma industries hesitate to adopt modern technologies that could boost their productivity. All biopharmaceutical companies need to have advanced and robust cybersecurity systems in place, as these systems play a crucial role in enabling companies to adopt newer, more innovative technologies [15, 16].

The challenges facing biopharmaceutical manufacturing in the last few years are varied, from formulation and delivery challenges to skilled labor shortages and shortened timelines for drug development and commercialization. The best possible solution for these hurdles is the integration of advanced technologies. Biopharma is accelerating its digital industry 4.0 transformation in pursuit of better cGMP (current good manufacturing practices) compliance. The pressure to adopt advanced information and data technologies across the enterprise will not abate, and the digitalization trend will have a significant impact on operational expenses and cost-of-goods accounting for years to come [17].

2 Biopharma Landscape: Gender Diversity

Biopharmaceutical companies are constantly growing, evolving, and developing innovative approaches to tackle complex diseases. The biopharma sector is seeing a wave of transformation on all fronts. Most biopharma companies are welcoming diversity, both in terms of people's identities and skill sets, thus creating a favorable environment for their employees to grow—both professionally and personally. Organizations are largely focused on optimizing functional capabilities and leadership development. One key success factor is developing a leadership team with greater diversity. Diverse teams make better decisions, both as individuals and collectively as a team, increase efficiencies, and exhibit more creative thinking [18].

From the beginning of the era of modern biotechnology, women have made groundbreaking contributions to the sector of life sciences and their applications. Yet, despite their key role in medical discoveries, the stories of many renowned female scientists remain largely untold, and their achievements have often been overlooked by their male colleagues in terms of patents, securing high positions in industry or academia, and becoming Nobel laureates [19]. Presently, women constitute about 50% of the talent pool in biopharma and hold more than half of PhDs. However, this number drops off abruptly at the leadership level with less than 10%

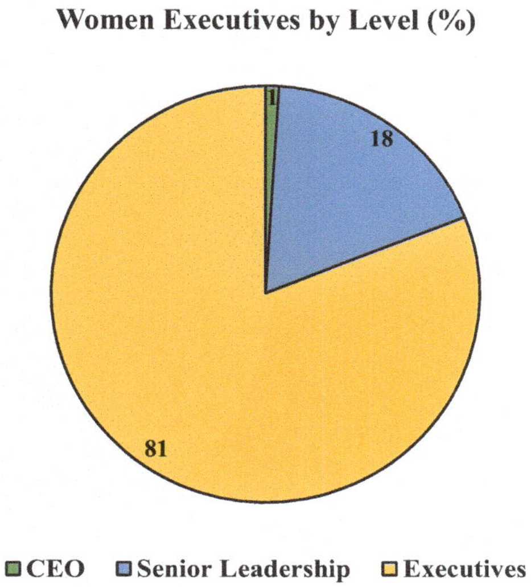

Fig. 1 Percentage of women executives by level in biopharma companies. (Source: Heidrick & Struggles Report on Biopharma Focus: Gender Diversity)

of the highest-valued biotech firms having females at the Chief Executive Officer (CEO) level. At the current rate, women are a "minority" in leadership but the majority in the highly skilled workforce in biopharma companies [20] (Fig. 1).

Gender diversity plays a significant role in big and small biopharma companies and now the share of women in senior management roles in most biopharma organizations worldwide is getting largely aligned with the location of the company. As per Heidrick & Struggles report, the United States represents 55% of women leaders in biopharma companies. European countries like France, Germany, Switzerland, and the United Kingdom also have a large share of female leaders in biopharma firms as compared to the Asia-Pacific regions [21].

2.1 The Position of Gender Diversity in Biopharma Today

According to the 2021 Business Insider report, it was observed that out of 182 biotechnology and pharmaceutical firms, only 8% had female CEOs (Fig. 2). The leadership teams in biotechnology and pharma companies are still mostly male. Only 21% of women make up biotech leadership teams and hold just 10% of the biotech board seats. This gender gap is concerning, as biopharma companies are one of the most powerful and fastest-growing industries, affecting the lives of billions of people worldwide daily. Comprehensive research has shown that the impact of gender diversity on organizations translates into better business performance like increased levels of innovation, improved talent attraction and retention, better return on investment, and reduced volatility; but also enhances risk mitigation strategies, improved competitiveness, more effective stakeholder management, and greater access to capital [22].

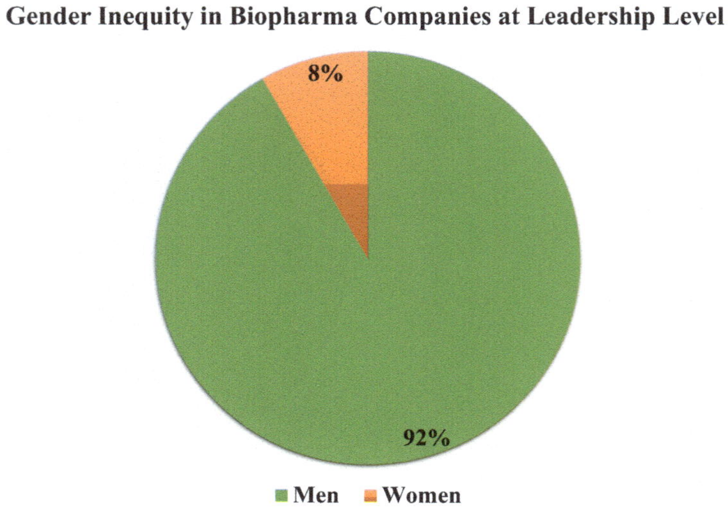

Fig. 2 Gender inequity in biopharma companies at the leadership level. (Source: Business Insider Report 2021)

As per the Heidrick & Struggles report, 70% of the women leaders were hired internally at the largest biopharma companies as compared to 57% at smaller companies. This survey showed that large companies are better at developing and retaining female leaders because of Diversity & Inclusion (D&I) programs and more opportunities for growth. Most of the biopharma companies have signified their commitment to diversity, equity, and inclusion (DEI), particularly with respect to gender equality, with more and more women in senior management positions [20]. For big biopharma companies, DEI strategies have become an integral part and pillar of ethical principles, resulting in fundamental values that are transforming into strategic business imperatives. Most biopharma organizations like Eli Lilly, MSD Italia, and Pfizer are trying to build a truly diverse and inclusive culture to develop diverse talents that can contribute to creating an atmosphere where people can express their full potential and thrive to contribute to the organization's success. Large European companies have facilities in many countries, presenting more chances for women to rise [21, 22]. As an example, two-thirds of MSD Italia's board members and 50% of their leadership team and managers are women, while 63% of new hires and promotions are women [23]. Additionally, this company is one of the first companies to eliminate the gender pay gap. Similarly, at Sanofi, Spain, women make up more than 53% of the organization [24].

2.2 *Women in Leadership Positions: Present Scenario*

In recent decades, there have been numerous calls for greater gender diversity in order to close the gender wage gap. Many strides have been made in this direction, but there is still a fundamental lack of women in leadership roles. Women hold

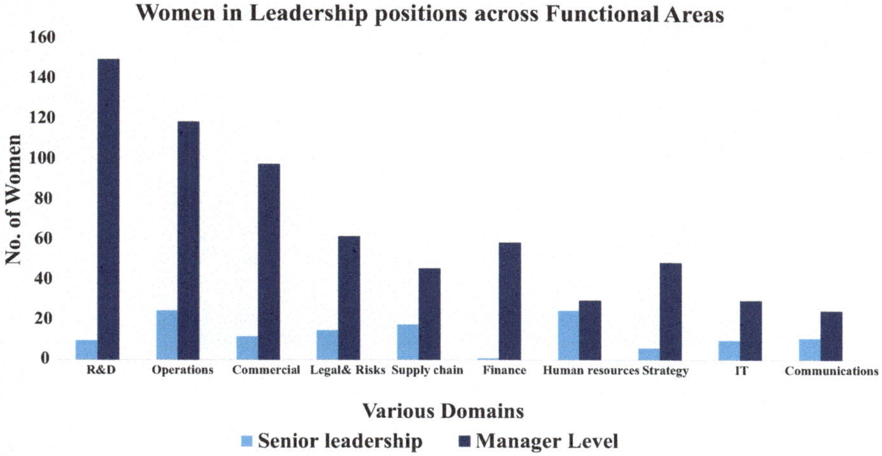

Fig. 3 Women in leadership positions across functional areas. (Source: Heidrick & Struggles Report on Biopharma Focus: Gender Diversity)

leadership roles across various domains and are prominent in numerous R&D functions that are core to pharma and biopharma operations. Figure 3 represents that in the senior leadership role, very few women are in the R&D domains compared with support functions and further in the other operational functions comprising strategy and supply chain [20, 25].

2.3 *Women Leaders Transforming Biopharma in India*

Women in India are gradually making their leadership presence in various fields like entrepreneurship, administration, education, engineering, health, etc., at regional, national, and global levels. As per the Economic Times Report 2022, Indian women hold 17.1% of board seats. Women are constantly evolving and reaching new milestones in the biopharma industries in India. The Indian biopharmaceutical sector has long been dominated by men, but now substantial change is expansive across this vastly technical and demanding business outlook.

There are a number of powerful women in the biotech industry in India who are constantly producing game-changing ideas and services in some of India's top biopharma companies. These women are making significant contributions to R&D and innovation in the field and are leading the way for the next generation of female scientists and professionals. In addition, these women have added tremendous value in framing their organizations [25]. Here are some of the most powerful women in the biotech industry in India (Fig. 4).

Kiran Mazumdar Shaw: She is the Founder and Executive Chairperson of Biocon Limited & Biocon Biologics, one of India's largest biotech companies. She is known for her innovative work in the biotechnology sector and has earned numerous

Fig. 4 Women leaders in indian biotech sector

national awards, including the Padma Shri (1989) and the Padma Bhushan (2005) from the Government of India (GOI) [26].

Dr. Renu Swarup: She is an Indian geneticist and former secretary of the GOI, previously heading the Department of Biotechnology (DBT) under the Ministry of Science and Technology. She has actively contributed to driving the growth and development of India's Biotechnology sector. She also contributed to the establishment of India's largest microbial resource center, the Microbial Culture Collection (MCC) in Pune, Maharashtra. She has implemented many innovative programs and policies to promote research, entrepreneurship, and collaboration in the biotechnology sector [26].

Dr. Soumya Swaminathan: She was the Chief Scientist at the World Health Organization (WHO), where she built the science division, focusing primarily on research, quality assurance of norms and standards, and digital health to improve health outcomes. She played a significant role during the COVID-19 pandemic by urging countries to conduct whole genome sequencing of the SARS-CoV-2 virus and upload the sequences to the Global Initiative on Sharing All Influenza Data (GISAID) project. She is also an Indian Pediatrician and clinical scientist known for her research on Human Immunodeficiency Virus (HIV) and Tuberculosis (TB) [26].

Dr. Preetha Reddy: She is the Vice Chairperson of Apollo Hospitals Group, one of India's biggest healthcare industries. She has played a significant role in the development and growth of the biotech division, which focuses on innovative therapies for a range of diseases [26].

Dr. Swati Piramal: She is the vice chairperson of the Piramal group and is an Indian scientist and industrialist. She is involved in healthcare, focusing primarily on innovation and public health, and has launched public health campaigns against chronic diseases, osteoporosis, malaria, tuberculosis, epilepsy, and polio [26].

2.4 *Challenges Faced by Women in the Biopharma Industries*

The majority of science, technology, engineering, and mathematics (STEM) professionals are mostly males. As per the World Economic Forum Report 2021, 33% of women are researchers and have attained equality when it comes to numbers in life sciences in many countries. Women comprise only 28% of graduates in engineering, and 40% of those are in computer sciences. Women account for just 22% of Artificial Intelligence (AI) employees globally. But many countries, like Singapore, Italy, and South Africa, are leading the way in driving diversity. In all three countries, women make up about 28% of AI professionals, compared to just 14% in Brazil, 15% in Mexico, and 16% in both Germany and Poland.

Women make up less than one-quarter of people engaged in STEM careers. The number of women entering the biopharma industry has increased in the past several decades, but female professionals continue to face many workplace challenges. The percentage of women in leadership roles in the biopharmaceutical sector is dismally low. The biopharma industry continues to struggle to make improvements in gender diversity.

The challenges faced by women in the biopharma industry can be classified into three main categories, namely family responsibilities, workplace environment, and stereotyping (Fig. 5) [27].

2.4.1 Family Responsibilities

Women who are married and have children face a higher level of family conflict than their male counterparts due to work and family pressure (marriage, childcare, and household activities), which negatively impacts career accomplishments. Women often experience a permanent lowering of their salary levels and benefits

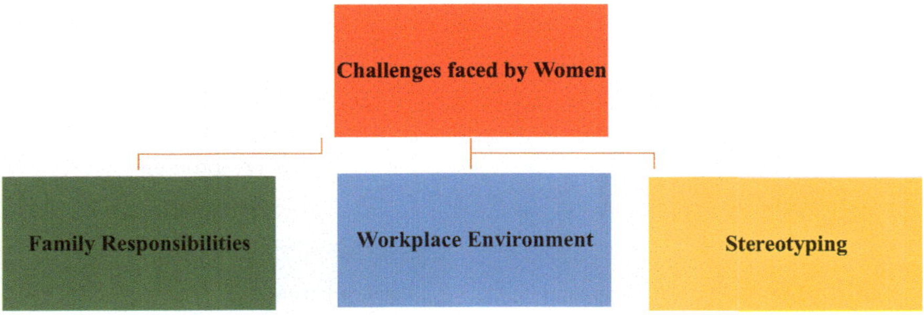

Fig. 5 Challenges faced by women in biopharma company

when they take maternity leave or time off to raise children early in their careers. Also, women are more pressured to choose family versus advancing their careers as compared to men. Globally, women continue to be paid less as compared to men for comparable work. As per the World Economic Forum report, for every dollar a man earns, on average, women are paid 54 cents, and it has been predicted that based on today's progress, it will take ~202 years for this gap to close. This gender inequality marks nearly all aspects of women's livelihoods, career opportunities, promotions, and advancements [27, 28].

2.4.2 Workplace Environment

The presence of barriers, such as inappropriate work schedules, prevents women from working and getting jobs. This results in a substantial negative effect on female motivation to work effectively. In addition, women face major obstacles to achieving fairness and equal opportunities in leadership because some male managers deliberately disregard and misbehave badly with female employees, which, in turn, leads to a stressful environment in the workplace. The challenges faced by women are listed below in Fig. 6 [27, 29].

(a) Pay Disparity: As per Pay Scale 2021 survey, women make $0.82 for every dollar a man makes. Over the last two decades, women have made great strides in opting for jobs that were previously performed exclusively by men. Additionally, women face discrimination in respect to recruitment, salary hikes, promotions, and many more. According to the World Economic Forum reports, the gender pay gap will not close for another 99.5 years [30].
(b) Sexual Discrimination: Sexual harassment in the workplace is one of the most common problems that women face. The #MeToo movement has played a vital role in bringing out numerous cases of women facing sexual and nonsexual harassment in the workplace. Thirty-five percent of women in full-time corporate sector jobs have experienced sexual harassment, as per the Women in the Workplace report. The Equal Employment Opportunity Commission (EEOC)

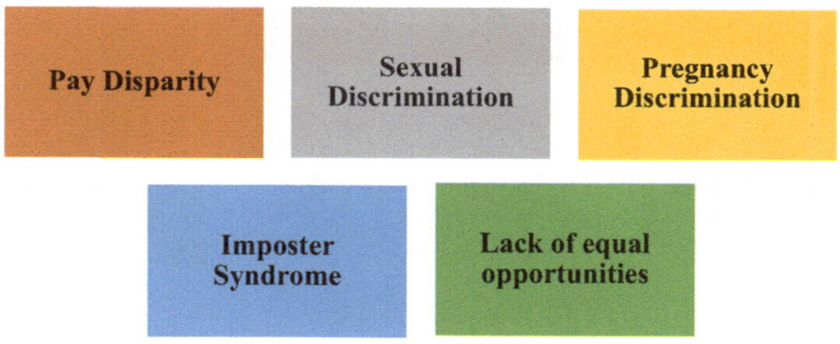

Fig. 6 Challenges faced by women at workplace environment

found that 75% of women subjected to hostile situations will not report their harassment, especially when the abuser is someone in a senior position [31].

(c) Pregnancy Discrimination: One major dilemma facing working women is family planning. As per Guardians report, more than 50,000 women lose their jobs over maternity discrimination in the United States. It continues to limit women's opportunities and economic advancement. Additionally, pregnancy discrimination in the workplace not only negatively impacts the mother but also negatively impacts the baby she is carrying while experiencing the discrimination [32].

(d) Imposter Syndrome: As per the KPMG study report 2020, it is estimated that ~75% of female executives across industries have faced imposter syndrome, while ~47% of female executives were in disbelief about reaching their level of success [33].

(e) Lack of equal opportunities: In most organizations, men are anxious about having a female boss. In managerial and C-suite roles, women are still underrepresented. As per the Deloitte Global Women in Boardroom Report—2022, it is estimated that an average of 19.7% of board seats are held by women worldwide as compared to India, where only 3.6% of women hold board chair positions [34].

2.4.3 Stereotyping

For many centuries, society has been forming gender stereotypes, conveyed through the media, which promote gender inequality. Gender stereotyping remains a persistent challenge within the biopharma industry. It leads women to feel demoralized and demotivated in their workplace, and this, in turn, increases the gender gap in STEM-related fields. The impact of gender stereotypes on women can be devastating [27]. For example, gender-specific behavior negatively affects the available opportunities for the career progression of women, resulting in constant minimal representation and participation of women in top-level management positions. In addition, pessimistic beliefs about women's accomplishments and success may harm their aspirations for career advancement. As a result, women may choose not to apply for leadership roles if they fear that they lack the ability to perform such roles [35].

2.5 Future Prospective: How to Increase Gender Diversity

In today's active business environment, gender diversity plays a significant role in the competitiveness and growth of any organization. Diversity and inclusion have become imperative in organizations and are powerful tools to drive innovation and growth in a work environment. A diverse and inclusive workplace with a range of approaches and perspectives allows for distinct ideas and creativity, which further improves performance, processes, and productivity. Gender diversity plays a crucial role in workplaces as it brings various viewpoints, creates more chances for equal representation, improves the overall quality of life for the employees, and increases employee satisfaction [36].

2.5.1 Benefits of Gender Diversity in the Workplace

Gender diversity is not only a fundamental human right but also essential for a peaceful, prosperous, and sustainable world. The main advantages of gender diversity in the workplace are listed Fig. 7 below:

(a) Enhanced Collaboration and Productivity: A gender-diverse company is more competitive as compared to ones that are not because women help develop team processes and promote collaboration. As per research conducted by Harvard, it was observed that gender diversity results in more productive companies, as measured by market value and revenue [37].
(b) Creates a Safer and Sensitive Work Environment: Companies that foster a fair and equitable balance between all genders ensure that the work environment is healthy. Every organization should work on creating stringent policies toward employee misconduct in the workplace like unethical relationships, harassment and discrimination, bullying, and racism. Implementing previously mentioned policies will help employees to bring their best selves to work; collaboration becomes easier, and employees do not feel discrimination in the workplace [38].

Fig. 7 Benefits of gender diversity in the workplace

(c) Flexibility: Women across the world do most of the voluntary caring. Women with children who can take advantage of flexible working hours are better able to maintain their career momentum and progress to the C-suite level or senior management level [37].
(d) Increased Profit: According to the McKinsey Report, it was found that organizations with higher gender diversity are 21% more likely to experience above-average profitability. Also, companies that have mixed-gender environments and diverse workforces were able to achieve 33% better-than-average profits [39].
(e) More Innovation: In a diverse workplace, employees will have various views and opinions because of various abilities and experiences. This will be incredibly useful for creating novel opportunities and outlining and accomplishing business strategies [38].
(f) Diverse Talent Pool: Organizations that lack women miss out on half of the population's expertise and skills. The benefits of a diverse talent pool include new creativity, increased productivity, and innovation in the workplace [37].

3 Careers in Biopharmaceutical Field: A Personal Journey of Dr. Rozaleen Dash and Ms. Ankita Bhargava

3.1 Personal Story of Dr. Rozaleen Dash

I am proud to be part of a field that is revolutionizing healthcare while fostering greater gender inclusivity. India is experiencing a significant surge in women's empowerment, with more women breaking barriers and taking on leadership roles across STEM fields, including biotechnology and biopharma. While historical challenges such as gender bias and limited representation persist, women are reshaping the narrative through resilience, innovation, and scientific excellence. My journey in this field has been fueled by determination and a passion for research, proving that women can lead ground-breaking advancements while balancing multiple roles in society. I was fortunate to grow up in a supportive family that provided me with equal opportunities for education and personal growth. From an early age, I was determined to challenge societal norms and demonstrate that women can excel beyond traditional roles, shaping careers that make meaningful contributions to society. My journey in science has been one of perseverance, creativity, and resilience, navigating challenges that often accompany being a woman in a male-dominated field.

Even in scientific research, gender disparities persist, with women constituting only one-third of researchers globally. Establishing myself in this field required immense dedication, particularly while balancing my professional responsibilities with my personal life as a wife and mother. Finding harmony between these roles has been both a challenge and a privilege, allowing me to embrace my career and personal commitments with equal dedication. My passion for science was fueled by curiosity from a young age—I often pondered questions like "Why do we dream?" and "Can we live forever?" With a deep fascination for the natural world, I pursued a Bachelor of Science (BSc Honors) in Zoology from Utkal University, India, followed by a Master's in Science in Zoology. My academic journey allowed me to explore the intersection of biology and research, leading to diverse career opportunities. However, after completing my master's degree, I took a career break of nearly four years to focus on my family and raise my beloved daughter. Returning to academia after this gap was daunting, but with unwavering support from my family, I persevered and resumed my scientific pursuits.

I embarked on my PhD journey at the All India Institute of Medical Sciences (AIIMS), New Delhi, under the mentorship of a distinguished researcher in immunology. My research focused on the expression of complement receptors 1 and 2 in relation to the pathophysiology and severity of Rheumatoid Arthritis (RA), a chronic autoimmune disorder. My research aimed to provide insights into the role of these receptors in human RA, contributing to a deeper understanding of disease progression and prognosis. Throughout this journey, I developed a systematic and analytical approach to research, reinforcing my commitment to scientific discovery. My passion for RA research was reflected in my contributions to the field, including multiple published papers. During my PhD, I secured my first independent research grant as a Principal Investigator from the Indian Council of Medical Research (ICMR), India—an achievement that strengthened my confidence as a researcher. Following my PhD, I joined the Centre of Excellence for Biopharmaceutical Technology (COE-CBT) at IIT Delhi as a postdoctoral research fellow, working under the guidance of a globally renowned scientist in the biopharma sector. My research focused on the characterization of monoclonal antibody (mAb) biotherapeutics and the functional comparability of biosimilars, along with evaluating critical quality attributes that influence drug potency.

In 2018, I was awarded my second independent fellowship grant (ICMR-RA) for my dream project, "Role of Inflammatory Cytokines on Aggregated mAb Therapeutics." I also played a key role in establishing the bioassay division at COE-CBT, IIT Delhi, demonstrating my expertise as an independent researcher and subject matter expert. My tenure at IIT Delhi has enriched my skills in functional characterization techniques for mAbs and biotherapeutics, enabling me to contribute significantly to drug potency evaluations. Beyond research, I have been actively involved in academic and industry collaborations. I have actively organized and coordinated numerous conferences and workshops, serving as a core member of the organizing team for the CBT Course Series from 2016 to the present. This initiative has attracted over 500 participants from academia and industry worldwide. In 2018, I was honored to be the Training Coordinator for the Global Initiative of Academic Networks (GIAN) at IIT Delhi, further expanding my engagement with the scientific community.

Additionally, as the coordinator for the Drug Development and Advanced Immunology courses at IIT Delhi, I have had the privilege of working with leading immunologists and experts in biotherapeutics and vaccine development. These collaborations have provided me with invaluable insights and strengthened my passion for research. I have also disseminated my findings at prestigious national and international conferences, such as the Gordon Research Conference (GRC) on Biotherapeutics & Vaccine Development, the European Society of Biochemical Engineering Sciences, the American Chemical Society, Bioprocessing Asia, and the Federation of Immunological Societies of Asia Oceania.

Growing up in an environment with limited career opportunities, I learned early on that perseverance, patience, and hard work are essential for success. Through years of training at India's leading institutes, I have gained invaluable expertise that fuels my commitment to scientific discovery and innovation. Today, as a woman in the biopharmaceutical sector, I am proud to be part of a new era of change-one where women are shaping the future of science and medicine. I am confident in my ability to contribute meaningfully to society by driving research that enhances healthcare and improves quality of life. My journey has not only strengthened my scientific acumen but also reinforced my belief that women can lead, innovate, and transform the scientific landscape.

3.2 *Personal Story of Ms. Ankita Bhargava*

As a woman in the biopharmaceutical sector, I take immense pride in being part of India's ongoing transformation. The momentum behind women's empowerment is growing across the country, with women now holding influential roles in various industries, including biotechnology and biopharma. It is inspiring to see the gender gap in technical education narrowing, particularly in biotechnology. Witnessing so many women excel in this field fills me with pride and fuels my determination to continue pushing boundaries. My fascination with biotechnology took root early, sparked by a deep curiosity about biology and its practical applications. This passion led me to pursue a Bachelor of Technology (B.Tech) in Biotechnology. During my undergraduate studies, I trained at Dabur Research Foundation, India, where I worked in the Quality Control department. This hands-on experience strengthened my understanding of biotechnology's practical applications and laid the foundation for my career.

Driven by a passion for continuous learning, I pursued a Master of Technology (M.Tech) in Biotechnology. During this time, my dissertation at the Serum Institute of India in Pune (India) provided me with extensive knowledge of vaccine immunology. After earning my M.Tech, I ventured into academia as a lecturer at the Noida Institute of Engineering and Technology (NIET), India, where I taught B.Tech students subjects such as cellular and molecular biology, immunology, and biosciences. This experience was incredibly rewarding, allowing me to mentor future biotechnologists while further refining my knowledge. Seeking new challenges, I transitioned to the Central Board of Secondary Education (CBSE) in Delhi as a Research Fellow (Healthcare Science). This role introduced me to curriculum development for healthcare education, workshop organization, and awareness programs for senior secondary students. Though a different field, it was fulfilling to contribute to healthcare education initiatives.

To gain regulatory experience, I joined the Biosafety Support Unit (BSU), the back office of the Review Committee on Genetic Manipulation (RCGM) under the Department of Biotechnology (DBT), India. As a Technical Assistant, I worked with DBT scientists, assisting in the development of draft guidelines and protocols for generating biosafety data to address challenges in emerging areas of biotechnology, biologicals, pharmaceuticals, and drugs. Additionally, I was involved in scrutinizing applications for biotechnology and pharmaceutical research, monitoring field and laboratory trials, and streamlining regulatory functions. This experience sparked my interest in regulatory affairs, a critical aspect of biopharma. Subsequently, I joined the Centre of Excellence for Biopharmaceutical Technology (CBT) at the Indian Institute of Technology (IIT) Delhi, (India) where I gained experience in project and event management and authored review articles and book chapters focused on biosimilar regulations in India and globally.

In 2020, I joined Gennova Biopharmaceuticals Ltd. (GBL) in Pune (India) as an Assistant Manager in the Project Management Department. This role was pivotal in my career, involving project management, regulatory operations , and coordination of key meetings. I later transitioned to the Business Development Department as a Manager, where I now manage DBT-funded projects, contribute to business development and investment strategies, and organize company-sponsored events both domestically and internationally. Additionally, I work closely with the Regulatory Affairs team to organize Institutional Biosafety Committee (IBSC) meetings and handle regulatory submissions on the RCGM portal

Reflecting on my journey, I am truly grateful for the diverse opportunities I have experienced across various domains. The biopharmaceutical industry is fast-paced and ever-evolving, demanding adaptability, technical expertise, and resilience. Navigating this field has been both challenging and fulfilling, and I remain committed to contributing meaningfully to its growth. Biopharma plays a transformative role in advancing scientific breakthroughs and improving global healthcare. It is an industry that fosters both professional and personal growth, and I am excited to continue making a difference. Looking ahead, I feel optimistic and driven, with a clear sense of purpose, eager to contribute to shaping the future of biopharma.

4 Make Biopharma a Rewarding and Meaningful Career Path

The biopharmaceutical industry has grown at an accelerated pace over the past few years. With new innovations and discoveries constantly evolving in the field of biotechnology, the purview of biopharmaceutical careers is increasing every day. There are countless benefits, which is one reason this career path has taken off exponentially.

Different skills and qualifications are accommodated here, and many people with different skill sets can find a place in a pharma organization. The options and opportunities are endless. The field of drug discovery, development, and commercialization offers a wealth of job and career opportunities for talented graduates with a sound background in life sciences or other academic disciplines and a genuine interest and motivation to start a professional career in the biopharmaceutical sector [40, 41].

4.1 Reasons to Choose Biopharma as a Career

In the biopharmaceutical industry, like every other industry, the pay depends on the level of qualification and skills. Better opportunities will be provided only if you acquire more knowledge. Even entry-level jobs in biopharmaceutical careers pay exceptionally well. Another benefit of pursuing a career path in biopharma is the opportunity to gain experience and discover new things. There are countless research projects going on in the biopharmaceutical industries. Being a part of finding new research is extremely rewarding because you know that what you do is going to help improve millions of lives all over the world [42]. One major reason to consider choosing a biopharmaceutical career is job satisfaction. There is a misconception that the work culture in the biopharmaceutical industry is rigid, as a stringent timeline is followed here to avoid any kind of error. However, there are a lot of people working here and many opportunities for flexible work hours. There are several new methods implemented in biopharmaceutical companies to ensure that human error is minimal. Every time the world is faced with a disease outbreak, the biopharma industry quickly switches up the workflow and provides the world of medicine with breakthroughs.

The biopharma industry operates in a diverse range of countries like the United States, the United Kingdom, France, Switzerland, Pakistan, India, Egypt, and more. A job in a biopharmaceutical company is an excellent chance to gain knowledge and experience about how various parts of the world operate. This industry also shares a diverse space with scientists, sales representatives, data analysts, and more. Working in the biopharma sector allows me to stay up to date with all the latest technological developments [42, 43]. The discovery of new drugs and chemicals is a major part of the industry. Many of the top biopharmaceutical industries in the world are known for manufacturing devices, such as chromatography equipment, bioprocessing containers, sterilizers, centrifuges, incubators, biosafety cabinets, bioreactors, and cell culture systems, which use the latest technology to improve the quality of life for millions of people. This industry is powerful and always working toward discovering new drugs and technologies that can be helpful for people [44].

4.2 Why Women Should Pursue Careers in Biopharmaceutical Sectors

Gender diversity is very prevalent across all industries including the biopharma industry, where women are underrepresented. In India, out of the 49.5 million-strong pharma workforce, only 11% are women.

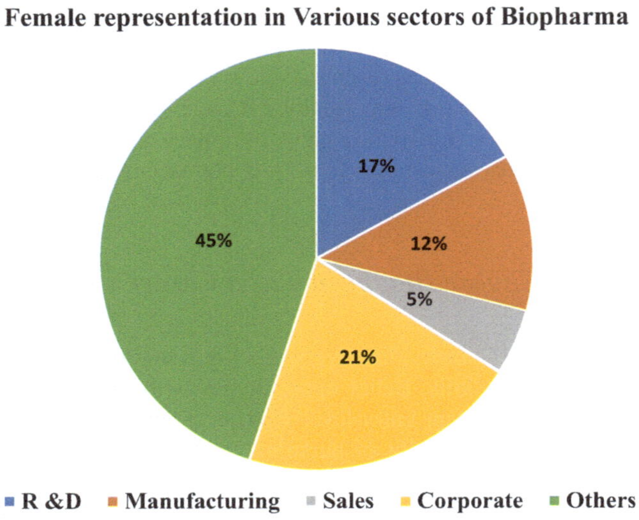

Fig. 8 Female representation in various sectors of biopharma. (Source: https://timesofindia.indiatimes.com/blogs/voices/how-women-are-bridging-the-gap-in-the-pharmaceutical-industry/)

The statistics will vary depending on the various sectors. R&D has 17% female representation, manufacturing has 12%, sales and marketing has a mere 5%, while corporate functions stand at 21% globally (Fig. 8). The figures are more dismal when it comes to female leadership roles. There is no reason to believe that biopharma is not suitable for women. It might be the case that, when the industry was established, women did not have the opportunity to work in otherwise stereotyped industries such as biopharma.

There are diverse career options for women depending on their skills and qualifications. Here are several reasons why a career in the biopharmaceutical industry can be rewarding for women [45–47]:

- There is tremendous work flexibility
- Skilled workers are needed every shift of every day
- High degree of job stability
- Opportunity to help people in the same medical community [48, 49].

In recent years, venture capital investments and significant contributions from existing biopharmaceutical companies to support biotech start-ups have been reaching new heights. Venture capital has something that is generally called a boys' club, but there are women making their mark in this area as well, including Kyparissia Sirinakis (the co-founder and managing partner at Epidarex Capital), Sara Nayeem (partner at New Enterprise Associates), and Arti Santhanam (executive director of the Maryland Innovation Initiative), to name a few.

Today, more and more women are making their mark in various sectors of the biopharma industry and breaking the conventional norms with continued efforts and equal representation in the workforce. Recently, biopharma companies all over the world, including India, are gearing up to transform their organizations to be more gender-neutral, with more women taking up leadership roles. Though the industry has some prominent women leaders, such as Zahabiya Khorakiwala, Managing Director at Wockhardt Hospitals; Kiran Mazumdar Shaw, Founder and Chairperson of Biocon Limited; Samina Vaziralli, Executive Director of Cipla; and Anjula Masurkar, Clinical Director at ENTOD Pharmaceuticals, a lot still needs to be done to address the gender gap issues in the Indian pharmaceutical industry. Women provide healthy leadership in some of the biggest biopharma companies in the industry. GSK, Viatris, Biocon, and Biological E all currently have female CEOs, and women make up just under 30% of executive directors at the top firms [50]. The reason is that similar numbers of men and women enter the workforce with advanced degrees in life sciences and medicine. For better decision-making, gender equity is needed across all biopharma industries.

4.3 Key Takeaway

The bottom line: Biopharmaceutical careers are highly lucrative, and there are countless opportunities for growth and learning. People working in the biopharma sector get numerous benefits, as well as exceptional salaries and high job satisfaction (Fig. 9). The biopharma industry is diverse, and there is a place for everyone [51].

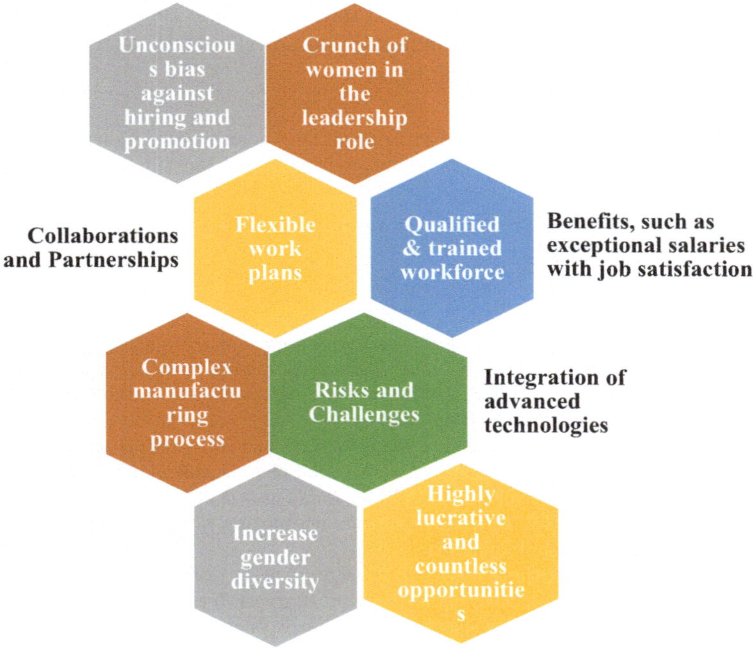

Fig. 9 Biopharma: a rewarding and meaningful career path for women

5 Conclusion

The biopharma industry continues to struggle to make progress with gender diversity. Though female representation remains low, many women leaders are making strides not only in their biopharma scientific fields but also in leveling the allegorical field to ensure that more women can follow in their footsteps in the decades to come. A few names who are in high-profile, influential positions in the biopharma business are Alice Zhang, Co-founder and CEO of Verge Genomics; Anat Ashkenazi, Chief Financial Officer of Eli Lilly & Company; Aviv Regev, Executive Vice President at Roche; Erandi De Silva, Co-founder and Senior Vice President of Forge Biologics; Janet Franklin, Vice president, global development Biosimilars at Amgen; Kiran Mazumdar-Shaw, Executive Chairperson and founder of Biocon Limited and Biocon Biologics Limited; and many more.

The biotech industry is still experiencing a lack of women in leadership positions. One major barrier is unconscious bias against hiring and promoting women in this industry. The focus should be on talent management, which plays a clear role in promoting equivalent numbers of men and women. A flexible work plan should be encouraged, especially for women returning to part-time roles after having children. Senior biopharma leaders should have strategies that encourage women to make an impact and enable them to reach the highest levels in their respective areas.

For real change to happen, industry executives and venture capitalists, primarily men, will need to go outside of their comfort zones. Another major focus should be on a mission that can help to develop a pipeline of diverse women in the biopharma sector through educational programs, paid internships, and funding academic science and accessibility. Finally, an inclusive culture should be encouraged where women feel welcome, respected, and heard.

Acknowledgments The authors would like to thank Dr. Vadiraja Bhat, Agilent Technologies, India, for trusting us and giving us this opportunity to connect with the editors.

References

1. Research & Markets. (2023). *Global biopharmaceutical market (2023–2028) by product type, therapeutic application, and geography, competitive analysis, impact of COVID-19 with ansoff analysis*. https://www.researchandmarkets.com/reports/5397295/global-biopharmaceutical-market-2023-2028-by
2. McKinley, M. (2021). *What's the difference between pharmaceutical and biotechnology?* https://www.morganmckinley.com/article/difference-between-pharmaceutical-and-biotechnology
3. Otto, R., Santagostino, A., & Schrader, U. (2014). *Rapid growth in biopharma: Challenges and opportunities*. McKinsey & Company. https://www.mckinsey.com/industries/life-sciences/our-insights/rapid-growth-in-biopharma
4. Grimpe, C., Minssen, T., Nicholson Price II, W., & Stern, A. D. (2022). Will mRNA technology companies spawn innovation ecosystems? *Harvard Business Review*. https://hbr.org/2022/04/will-mrna-technology-companies-spawn-innovation-ecosystems

5. CPhI. (2021). *Pharma trends 2022* (pp. 6–7, 16) [online]. CPhI. Available at https://www.cphi-online.com/cphi-pharma-trends-2022-report. Accessed 10 Jan 2022.
6. FDA's Center for Drug Evaluation and Research. (2022). *Advancing health through innovation: New drug therapy approvals 2021* (pp. 11–120) [Internet]. US Food and Drug Administration (FDA). Available from: https://www.fda.gov/drugs/new-drugs-fda-cders
7. Bonabeau, E., Bodick, N., & Armstrong, R. W. (2008, March). A more rational approach to new-product development. *Harvard Business Review, 86*(3), 96–102, 134.
8. Peck, R. W. (2007). Driving earlier clinical attrition: If you want to find the needle, burn down the haystack. Considerations for biomarker development. *Drug Discovery Today, 12*(7–8), 289–294.
9. Kola, I., & Landis, J. (2004). Can the pharmaceutical industry reduce attrition rates? *Nature Reviews. Drug Discovery, 3*(8), 711–715.
10. Dunn, A., & Pagliarulo, N. (2019, May 28). *Follow the money: How biopharma CEOs and workers got paid in 2018.* https://www.biopharmadive.com/news/biotech-pharma-ceo-employee-pay/554283/
11. Danzon, P. M., Nicholson, S., & Pereira, N. S. (2005). Productivity in pharmaceutical-biotechnology R&D: The role of experience and alliances. *Journal of Health Economics, 24*(2), 317–339.
12. Woodcock, J. (2007). The prospects for "personalized medicine" in drug development and drug therapy. *Clinical Pharmacology and Therapeutics, 81*(2), 164–169.
13. Gregson, N., Sparrowhawk, K., Mauskopf, J., & Paul, J. (2005). Pricing medicines: Theory and practice, challenges, and opportunities. *Nature Reviews Drug Discovery, 4*(2), 121–130.
14. Garnier, J. P. (2008). Rebuilding the R&D engine in big pharma. *Harvard Business Review, 86*(5), 68–70, 72–6, 128.
15. Goodman, M. (2008). Market watch: Pharma industry strategic performance: 2007–2012E. *Nature Reviews Drug Discovery, 7*(12), 967.
16. Kramer, J. A., Sagartz, J. E., & Morris, D. L. (2007). The application of discovery toxicology and pathology towards the design of safer pharmaceutical lead candidates. *Nature Reviews Drug Discovery, 6*(8), 636–649.
17. Kola, I. (2008). The state of innovation in drug development. *Clinical Pharmacology and Therapeutics, 83*(2), 227–230.
18. Cole, B. M. (2020). 8 reasons why diversity and inclusion are essential to business success. *Forbes*. https://www.forbes.com/sites/biancamillercole/2020/09/15/8-reasons-why-diversity-and-inclusion-are-essential-to-business-success/?sh=3379cb131824
19. Masurka, A. (2022). *How are women bridging gap in pharma industry?* Express Pharma. https://www.expresspharma.in/how-women-are-bridging-gap-in-pharma-industry/
20. Veith, M. F., & Sood, I. (2020). *Biopharma focus: Gender diversity*. Heidrick & Struggles. https://www.heidrick.com/en/insights/biopharmaceuticals/biopharma_focus_gender_diversity
21. Koberstein, W. (2017). The women of biopharma – Will they gain or lose ground in 2018? *Life Science Reader*. https://www.lifescienceleader.com/doc/the-women-of-biopharma-will-they-gain-or-lose-ground-in-0001
22. Livingston, S., & Dunn, A. (2021). Men fill more than 9 in 10 biopharma CEO positions, an inequity that costs women more than $500 million in pay each year. *Business Insider India*. https://www.businessinsider.in/science/health/news/men-fill-more-than-9-in-10-biopharma-ceo-positions-an-inequity-that-costs-women-more-than-500-million-in-pay-each-year/articleshow/83989382.cms
23. MSD. (2021). *8 women who made major strides at our company and beyond*. https://www.msd.com/stories/8-women-who-made-major-strides-at-our-company-and-beyond/
24. Sanofi. (2022). *Sanofi achieves the third year of Gold-level Gender Parity Certification*. https://sanoficanada.mediaroom.com/2022-05-03-Sanofi-achieves-the-third-year-of-Gold-level-Gender-Parity-Certification
25. The Economic Times. (2016). *Meet India's next generation of women leaders in pharma*. https://economictimes.indiatimes.com/small-biz/entrepreneurship/

meet-indias-next-generation-of-women-leaders-in-pharma/articleshow/53273951.cms?utm_source=contentofinterest&utm_medium=text&utm_campaign=cppst
26. Biopatrika. (2022). *7 most influential women in Indian biotech*. https://www.linkedin.com/pulse/7-most-influential-women-indian-biotech-biopatrika/
27. ALobaid, A. M., Gosling, C. M. R., Khasawneh, E., McKenna, L., & Williams, B. (2020). Challenges faced by female healthcare professionals in the workforce: A scoping review. *Dovepress, 13*, 681–691. https://doi.org/10.2147/2FJMDH.S254922
28. UN Women. *Equal pay for work of equal value*. https://www.unwomen.org/en/news/in-focus/csw61/equal-pay
29. Stamarski, C. S., & Son Hing, L. S. (2015). Gender inequalities in the workplace: The effects of organizational structures, processes, practices, and decision makers' sexism. *Frontiers in Psychology*. https://doi.org/10.3389/2Ffpsyg.2015.01400
30. Aragão, C. (2023). *Gender pay gap in U.S. hasn't changed much in two decades*. https://www.pewresearch.org/short-reads/2023/03/01/gender-pay-gap-facts/
31. EEOC Data Highlight. *Sexual harassment in our nation's workplaces*. https://www.eeoc.gov/data/sexual-harassment-our-nations-workplaces
32. De Piero, G. 50,000 women lose jobs over maternity discrimination, study shows. *The Guardian*. https://www.theguardian.com/women-in-leadership/2013/nov/20/fifty-thousand-women-lose-jobs-maternity
33. KPMG. (2022). *KPMG study finds 75% of female executives across industries have experienced imposter syndrome in their careers*. https://info.kpmg.us/news-perspectives/people-culture/kpmg-study-finds-most-female-executives-experience-imposter-syndrome.html
34. Deloitte. (2020). *Deloitte Global's latest Women in the boardroom report reveals crucial link between women's leadership and more diverse boards; overall rate of progress remains slow*. https://www.deloitte.com/an/en/about/press-room/women-in-theboardroom-report.html
35. Ding, W. (2013). Stereotyping prime obstacle to women in commercial science: Cracking glass ceiling starts with academic leaders, tech transfer offices. *Science Daily*. https://www.sciencedaily.com/releases/2013/03/130312152043.htm
36. Line Zero. (2020). *The positive effects of gender diversity in the workplace*. https://www.line-zero.com/blog/positive-effects-gender-diversity-in-the-workplace
37. Vantage Circle. (2023). *7 benefits of promoting gender diversity in the workplace*. https://blog.vantagecircle.com/gender-diversity/
38. Line Zero. (2023). *The positive effects of gender diversity in the workplace*. https://www.line-zero.com/blog/positive-effects-gender-diversity-in-the-workplace
39. McKinsey & Company. (2020). *Diversity wins. How inclusion matters?* https://www.mckinsey.com/~/media/mckinsey/featured%20insights/diversi-ty%20and%20inclusion/diversity%20wins%20how%20inclusion%20matters/diversity-wins-how-inclusion-matters-vf.pdf
40. 10 reasons why the world needs more women in leadership roles. Published on: 2021, March. https://www.naturalhr.com/2021/03/23/10-reasons-why-the-world-needs-more-women-in-leadership-roles/
41. Cuatristas, P. (2006). Drug discovery in jeopardy. *The Journal of Clinical Investigation, 116*(11), 2837–2842.
42. DiMasi, J. A. (2002). The value of improving the productivity of the drug development process: Faster times and better decisions. *Pharmacoeconomics, 20*(Suppl 3), 1–10.
43. Aitken, M., Berndt, E. R., & Cutler, D. M. (2009). Prescription drug spending trends in the United States: Looking beyond the turning point. *Health Affairs (Millwood), 28*(1), 151–160.
44. Lichtenberg, F. R. (2005). The impact of new drug launches on longevity: Evidence from longitudinal, disease-level data from 52 countries, 1982–2001. *International Journal of Health Care Finance and Economics, 5*(1), 47–73.
45. Cohen, F. J. (2005). Macro trends in pharmaceutical innovation. *Nature Reviews Drug Discovery, 4*(1), 78–84.

46. Abou-Gharbia, M., & Childers, W. E. (2014). Discovery of innovative therapeutics: today's realities and tomorrow's vision. 2. Pharma's challenges and their commitment to innovation. *Journal of Medicinal Chemistry, 57*(13), 5525–5553.
47. Panmoli, F., Magazzini, L., & Riccione, M. (2011). The productivity crisis in pharmaceutical R&D. *Nature Reviews Drug Discovery, 10*(6), 428–438.
48. Hinchliffe, E. (2020, May 18). The number of female CEOs in the Fortune 500 hits an all-time record. *Fortune*.
49. Shimura, H., Masuda, S., & Kimura, H. (2014). Research and development productivity map: Visualization of industry status. *Journal of Clinical Pharmacy and Therapeutics, 39*(2), 175–180.
50. Grabowski, H. (2004). Are the economics of pharmaceutical research and development changing: Productivity, patents and political pressures. *Pharmacoeconomics, 22*(2 Suppl 2), 15–24.
51. Cook, J., Hunter, G., & Vernon, J. A. (2009). The future costs, risks and rewards of drug development: The economics of pharmacogenomics. *Pharmacoeconomics, 27*(5), 355–363.

Women with Advanced Degrees

Leyna Zhao

Contents

1 Building Foundation and Resilience

On June 21, 1997, as a 21-year-old college graduate, I traveled from Beijing, China to Columbus, Ohio. I was full of curiosity, eagerness, and hope. I was to start my graduate program in the chemistry department at the Ohio State University (OSU). I was determined to become a professor one day, just like my dad. I was ready. But was I?

The first 2 years of graduate school were rather simple. By the time we are in graduate school, we are good at taking classes and passing examinations. It is always the new "tricks" that take time to master. Standing in front of 25 college kids teaching a laboratory is new. Reading, digesting, and presenting journal articles in front of all department students and faculty is new. Taking cumulative examinations on all subjects of chemistry every first Saturday of the month or failing four times in a row is new. If you are in the midst of the same chaos, please take a deep breath. I want to reassure you that you will become very good at this, and the skills you build through this experience will benefit you for a lifetime. You can only further develop yourself when you are pushed outside your comfort zone. It is during these 2 years of intensive training that I started to master public speaking skills, became

L. Zhao (✉)
Product Management – Agilent Technologies, La Jolla, CA, USA
e-mail: Leyna.Zhao@agilent.com

S. Taylor, M. Campbell (eds.), *Women in Biopharma*, Women in Engineering and Science, https://doi.org/10.1007/978-3-031-80348-2_5

able to independently read and understand various research topics, and, most importantly, established the habit of reading broad research topics.

The next 4 years of graduate school seemed simple but were rather complex. I was fortunate to join Dr. Ming-Daw Tsai's laboratory. He is well known in the space of enzyme kinetics and protein structural and function relationships. His research projects are funded by three RO1 grants, which made his lab one of the biggest and most well-funded labs in the chemistry department at OSU. Each junior graduate student is paired with a senior graduate student on the same project. The junior graduate student assists the senior graduate student with their project and co-authors a paper initially. Once independent, the junior graduate students propose their own project, test their hypothesis with experiments, publish, and graduate. In between, they work with their advisor to write the RO1 grant.

Thinking back, I am full of gratitude to my advisor for giving me the chance to grow and teaching me how to be patient, how to trust, and how to be independent. However, as a fourth-year graduate student with no first-author paper, I was by no means calm or appreciative. I felt desperate, I was frustrated, and I sensed overwhelming pressure. When you are in a "sink-or-swim" situation, you work and think a hundred times harder. Once you learn to swim and survive, you realize that you can learn anything, and you can survive in any situation. Many of my Ph.D. colleagues have succeeded in many different leadership positions in academia and industry. I believe their successes are inseparable from the same Ph.D. training we had in Tsai's lab.

If you are in the same boat as I was, please give it a little more time and don't give up. Graduate school, in many ways, is meant to take away all our pride and make us into a new, humble person. That new, humble person will be independent and not afraid of tackling any problems. Independent thinking and problem-solving skills are what employers look for, regardless of whether you are applying for a position in academia or industry.

I graduated with multiple first-author papers in the top journals in the field (Journal of Biological Chemistry, Journal of the American Chemical Society and Biochemistry) and numerous abstracts, including podium presentations at the American Chemical Society Annual Meeting and Gordon Conferences. I was convinced I would follow an academic path. Therefore, I sought opportunities to expand my field of expertise from protein chemistry to cancer research. I purposely chose to go out of my comfort zone and applied for postdoctoral positions with many members of the National Academy of Sciences in California.

The next chapter of my life was a postdoctoral study with Dr. Peter K. Vogt at The Scripps Research Institute. Dr. Vogt is a world-renowned virologist who contributed to the discovery of the first oncogene—Proto-oncogene tyrosine-protein kinase SRC. His passion for science, life, art, and mentoring is so immense and inspiring. During the 5 years (2004–2009) I worked with Dr. Vogt and many other outstanding colleagues, I transitioned from a pure protein chemist to a cancer biologist. My brain was no longer confined to an enzymatic reaction in a highly controlled condition. I started to embrace the complexity of biology, signal transduction pathways, and translational biology leading to cancer therapeutics. I was able to build on my structural biology knowledge to identify the oncogenic mechanism of the hot-spot mutations in phosphatidylinositol 3-kinases (PI3K).

As the first step toward my future academic career, I wrote my first independent grant and was awarded the Ruth L. Kirschstein National Research Service Award for Postdoctoral Fellows from the National Institute of Health (1 F32 CA130304) from 2007 to 2009. In the meantime, I realized political situations could affect science so much. Despite the great science generated from Vogt's lab and the collaboration, Dr. Vogt's funding situation worsened. Many RO1 grants our lab tried to renew were rejected between 2008 and 2009. I was one of a few lucky postdocs who had our own funding and could therefore continue our research in the lab. I was very productive during this time: publishing seven peer-reviewed publications, including three first-author papers. I then had to make the most difficult, but sensible, decision of my career: to transition from academia to industry.

Job searching during the 2009 recession was hard. I took two routes: exercising my relationships and applying via job posting sites (e.g., BioSpace.com). Utilizing my relationships seemed to work initially as these connections led to quite a few interviews with big biopharma companies, but none of the interviews led to an offer. These big biopharma companies were looking for scientists with specific skill sets and I was not the perfect fit for these well-defined jobs. On the other hand, one of the jobs I applied to via BioSpace finally resulted in a phone interview with a small local biotech company, ACEA Bioscience. They were looking for a bench scientist who could conduct toxicity tests with cell-based assays. This is one of the many jobs I was qualified for. In preparation for the on-site job interview, I started my research on the company. I was fascinated by their proprietary impedance-based xCELLigence Real Time Cell Analysis technology and its capabilities. Trained as a biologist, "seeing is believing" is a part of me. I still remember challenging the founder and Chief Technology Officer (CTO), Dr. Xiaobo Wang, at the interview by asking, "but the biologists would naturally want to see the cells, is it possible to have a view window on the biosensor plate?" At that very moment, Xiaobo smiled, took out a beta product, and told me that the new biosensor plate with a view window was launching next month.

Thinking back, I know that I scored my interview by doing my homework on the company, so I was prepared to ask the right questions. Now that I am at the other end of the table interviewing job candidates, I am energized by the candidates who have done their homework about our company and ask relevant questions. If you are at the stage of interviewing for your next job, I would recommend that you treat your interview seriously, study the company, and prepare 1–2 questions that are refreshing to the interviewers.

2 Growing with the Job: Be Hungry, Be Humble, Work Hard and Be Ready for Change

I started my first industry career with ACEA Biosciences on November 1, 2009, as a bench scientist. The annual salary was in the low to mid $70K's which was a huge step up from my postdoc salary of the low $30K's. My initial project was identifying the Mechanism of Action (MOA) of a few screen hits and working with medicinal chemists to improve the in vitro efficacy. My manager gave me the freedom to utilize my solid biochemistry and cell biology experiences to tackle scientific problems. With so many parallel projects and experiments, I benefited significantly from ACEA's proprietary xCELLigence Real Time Cell Analysis technology. All my in vitro cell-based assays became simple, walkaway operations, where I could seed my cells on a Thursday, add compounds on Friday, and come back on Monday with all the results ready to present. I became extremely productive during my tenure as a bench scientist between 2009 and 2012. I felt truly liberated by the label-free, real-time technology. With all this newfound free time, I was able to think more, learn more, collaborate more, and accomplish more.

Here are a few examples. I would like to challenge the young readers of this book chapter. If the following situations happened to you, what would you do?

1. I identified a radioactive assay to confirm the MOA of a leading drug candidate. Unfortunately, ACEA didn't have the license to conduct radioactive assays. What would you do in this situation? Give up or seek a solution?
2. In pharma or biotech companies, if a discovery is not patented, it cannot be publicly disclosed. Would you just accept that as fact and go to a conference without the opportunity to present and possibly gain an investor's interest?
3. Resources are naturally limited in a small biotech company. There is a need to understand the binding mode of lead compounds with the target proteins. Outsourcing these activities is extremely costly. Do you simply leave the problem to your manager?

My answers to all these questions are no. In the first case, I reached out to my postdoc advisor at Scripps Research Institute, did my radioactive assay in his lab, and quickly confirmed that my lead drug candidate binds to tubulin in the same site as the FDA-approved drug Colchicine. In the second case, I proactively approached our CTO and drafted a patent, which allowed me to present my research at scientific conferences. For the third case, I proposed to my manager to allow me to take the lead on this project based on my structural biology background.

I learned and mastered a new commercial docking software (Schrödinger) during my 3-month maternity leave after giving birth to my second child. I then subsequently identified the mechanism of action of our lead compounds as the third-generation Epidermal Growth Factor Receptor (EGFR) inhibitor, which binds and forms a covalent bond with mutant EGFR (T790M).

Now, thinking back, the new opportunity and the next chapters of my career were tightly related to my learning mindset and the actions that followed. The year 2012 marked a turning point for ACEA as well as my career transition. Roche informed ACEA that they were going to discontinue a 5-year strategic collaboration, where Roche was ACEA's exclusive marketing and distribution partner. To survive this transition, ACEA quickly launched a noncompeting instrument (iCELLigence) in mid-2012. Every scientist was empowered to identify beta evaluators and new sales opportunities. I identified a beta evaluator at Scripps Research, where I conducted my postdoc training. I led the beta evaluation and subsequently sold two iCELLigence to this very lab. A few months later, my manager approached me and asked if I was willing to take on a marketing role. I did not know anything about marketing, but I was confident I could tackle any challenges. I said "Yes," then signed up for a UCSD extension class—"Marketing 101" the same day.

I am extremely grateful that in my career, I have met mentors who were willing to trust and empower me to try new things. Dr. Vogt hired me as a postdoc and transitioned me from biochemist to cell biologist. Now, Dr. Yama Abassi, who was my hiring manager and has been mentoring me for 2 years, took a leap of faith and gave me a position as a global marketing manager, trusting that I could quickly transition and morph into a marketer. I made a promise to myself that I would not let Yama down.

I was hungry, worked hard, and learned from everyone. I stayed customer-focused and took, "peer-to-peer marketing" to the next level. Between 2013 and 2016, I led the launch of many new products, new websites, and peer-to-peer marketing events, including eight user symposiums. As a company, we secured our position in the cancer immunotherapy and cardiotoxicity fields. Our business was going well, and I fully enjoyed my work. However, my health took a bad turn. Since the departure of the global support manager, I took on the responsibility of both support and marketing. I was totally burned out; I was admitted to the hospital for heart issues. In response to the situation, my manager quickly hired a marketing lead and assigned me to oversee only scientific support. I have to say that I was definitely not thrilled. But I didn't choose to do a mediocre job just because it was not my passion. I constantly asked myself, "what can I do to make a difference to the business?" I started to identify gaps, establish new processes, and seek opportunities.

In 2016, our commercial team used Salesforce to track leads. Unfortunately, our install base was still in a Microsoft Excel spreadsheet, and we didn't have a CRM (customer relationship management) system. Recognizing this was a huge gap, it also presented a giant opportunity to enhance the customer experience and extend the customer life cycle for repeat business, consumable sales, and service offerings. Since I had no additional resources that I could assign to the project, I decided to take on the challenge. I used personal time to manually input the install base into Salesforce, validating and updating the customer information in the Microsoft Excel file. Later, I was able to recruit other team members who believed in my vision to do this outside their job responsibilities. With joint efforts, the CRM system was established. I then established three processes to ensure the full return on investment of these huge efforts. (1) all new installation and warranty information are logged faithfully into the system; (2) all service cases are associated with the install base and logged in CRM; and (3) warranty renewal notifications are sent out 3 months prior to the expiration date. I also took the lead and drafted the first service brochure to promote our service offerings. With multipronged approaches, our service revenue started to double every year.

3 Conclusion

As I grew from a bench scientist to a global marketing manager and then to Director of global scientific support, I have always been true to my heart. I have been inspired by many leaders in the company; mirroring them and learning from the best have always been the development goals I set for myself.

I have met many renowned scientists and noble prize laurates. All share one striking characteristic: they were all humble. At this point, I must write a little more about another person who influenced my character greatly. Dr. Yama Abassi, who hired me at ACEA Bioscience and was my direct line manager for 6 years. I still vividly remember that we had a lunch team meeting and the trash went too full. Yama went over and took out the trash. I was shocked. It is this "no task is beneath me" and "do whatever it takes to keep the ship afloat" mentality that has greatly shaped how I also like to be.

ACEA was acquired by Agilent in November 2018. Working in a small private biotech company versus a publicly traded company was a huge change. Through the long journey of integration, I have met many great people who shared the mindset of "we want to grow the business, not break the business." This has encouraged a culture of "speaking up," which became one of Agilent's core values in 2023.

Fifteen years later, I am now leading a team of ten product managers and overseeing four product portfolios within Agilent's Cell Analysis division. I am thrilled by the vast opportunities that lie in front of us. I will continue to do my best to contribute to the business and to be a hungry, humble, work hard, and resilient leader who can inspire others. And I hope my story will shed some light on your career journey.

My Journey as a Woman in Biopharma: Embracing Challenges and Finding My Passion

Isadora Ventura-Quintanilha McCullough

Contents

1 Life in Brazil

To tell you my story as a woman in biopharma, I need to rewind to the very beginning of my journey in the United States. An audacious decision for a 17-year-old girl that forever altered the course of my life. Growing up in Rio de Janeiro, Brazil, was amazing. I always enjoyed Rio; it was a perfect mix of a big city and the beach (Fig. 1).

However, when I was a teenager and had to decide what career path I would take, I became very focused on my future, and I learned that being fluent in English could help me secure a great job. So, the summer before my senior year, I convinced my parents to allow me to come to California as an exchange student. I promised them that graduating from high school in the United States would allow me to learn the English language, and that, consequently, it would allow me to do well in my future

I. V.-Q. McCullough (✉)
Agilent Technologies, Santa Clara, CA, USA
e-mail: Isadora.mccullough@agilent.com

S. Taylor, M. Campbell (eds.), *Women in Biopharma*, Women in Engineering and Science, https://doi.org/10.1007/978-3-031-80348-2_6

Fig. 1 Author with husband and parents at Christ the Redeemer, Rio de Janeiro

career. As I had taken a few English classes, they didn't go as smoothly as I had hoped, so my English proficiency was limited to the most basic phrases. My parents were incredibly supportive of my decision, and together, we embarked on the extensive process of planning my move as an exchange student. They learned from a friend that her son had been an exchange student through Rotary, an organization with a global network dedicated to creating positive change. This friend's son had a wonderful experience, and my parents were drawn to what Rotary offered.

What stood out about Rotary was that, in addition to allowing me to go abroad, my parents would also need to accept exchange students into our home while I was away. Many other companies provided monetary compensation to host families, but Rotary's rules reassured my parents. They understood that families accepting Rotary exchange students did so not for financial gain but out of genuine care for the students' well-being. This assurance made them feel that I would be safe and well-cared for during my exchange.

We attended several Rotary meetings and learned about their selective process. I had to undergo a series of tests, including general knowledge, Rotary knowledge, and English proficiency. The results would determine our placement on a list, with the highest scores at the top. As openings became available, they would be offered to students starting from the top of the list. I dedicated a lot of time after school studying for these tests, determined to earn a high score.

2 Decisions

My top choices for exchange destinations were Australia, Canada, and New York, but there was no guarantee that these places would be available for the upcoming year. After completing the tests, we eagerly awaited news from Rotary regarding available positions. I was third in line on the list and anxiously anticipated the possibilities. The first email I received presented an opportunity in California. They provided me with a map of the territory, I had seen Hollywood in movies and imagined the vibrant culture of California, and I didn't hesitate to reply with a resounding "YES."

A month later, an email arrived from a woman named Tammy McBride. She shared details about her family and inquired if I was still interested in becoming their exchange student. Her family sounded incredible—I would have two host sisters, Victoria (12 years old) and Alexandria (Lexi) (16 years old), as well as two cats (Gigi and Bela) and one dog (Bummer). Tammy's mother, Judy, lived in a house behind theirs, and she was affectionately known as Grandma to everyone (Fig. 2). Tammy sent me photos of her family, their town, Capitola, and she shared heartwarming stories. It was evident how excited they were to welcome an exchange student into their lives.

And so, at the age of 17, I embarked on a life-changing journey as I moved to Santa Cruz, California, in the United States, to begin my adventure as an exchange student.

Fig. 2 Host grandma, author, and host mom

I was incredibly excited to be in California. The houses looked so different, and in Capitola, there were no tall apartment buildings. Everything seemed smaller, like a scene from a charming small-town movie. I arrived during summer break, which gave me a few weeks to explore with my host family. Everything was different from what I was used to back home. As the days went by, I felt a mix of excitement and loneliness. Not being fluent in English was tough. Communicating and expressing myself were challenging. I called my parents often, finding comfort in their voices. I exchanged emails with my friends a lot and felt a tinge of sadness missing out on all the fun they were having in their senior year.

In the weeks leading up to the start of school, all I had were my host family and the other exchange students. Rotary organized activities for all the exchange students within a 50-mile radius, and we became a close-knit group of friends (Fig. 3). We shared our homesickness and the excitement of being exchange students. When classes finally began, I was eager to meet new people and make friends. It was a bit challenging at first because my English was improving but still far from perfect. However, I pushed myself to say "hi" and introduce myself. Word spread that I was from Brazil, and some students were incredibly welcoming. I slowly formed some wonderful friendships, and they were incredibly helpful with both my classes and my accent.

Despite all the homesickness and challenges, my year in California was still great. I learned the language and fully enjoyed myself. I was lucky to be placed with a host family that was very patient with me and took great care of me. I felt at home, and for an only child, it was an adventure to suddenly have two little sisters. I

Fig. 3 Exchange students from Rotary 2017

learned a lot from my experience, created lifelong friends, immersed myself in the culture, and experienced a lot of personal growth. However, there were challenges along the way. Because I was a non-native speaker during my senior year of high school, school was sometimes terrifying. I had to deal with nicknames and laughter due to my strong accent. But I also felt accomplished and believed I could achieve anything I set my mind to.

3 Returning to Brazil Short-Term; Returning to the US Long-Term

After a fulfilling year abroad, I realized there was still so much more to learn in the United States. I decided to return to Rio de Janeiro for the summer and speak to my parents one more time, asking to move back to California to pursue my college degree. I mentioned that there was a community college close to where I stayed and that I would like to return to Santa Cruz to keep learning. Once again, they said yes, and once again, I packed up my belongings and came to California. For a few months, I stayed with my host family again until I found a room to rent in a house close to campus.

Back then, I aspired to graduate as a chemical engineer and to work at Petrobras, a petroleum company in Rio de Janeiro, a dream I had nurtured since high school. In my first year of high school in Brazil, something wonderful happened as I developed a real love for chemistry. I had an amazing teacher who inspired me and encouraged me to excel and explore further. Then, in my second year of high school, I took a career aptitude and personality test. It was an eye-opener! I realized that my true passion was in chemical engineering. I was captivated by the extensive problem-solving it entailed and the diverse array of career opportunities it offered. Chemical engineering is an exciting field that combines principles of chemistry, physics, and math to solve real-world problems related to chemicals and processes. This career could allow me to work in various industries, such as pharmaceuticals, food and beverages, energy, cosmetics, and more. For example, I could work to develop ways to produce medicines, design systems to purify drinking water, or create renewable energy sources.

Excited to embark on this journey, I enrolled at Cabrillo College, a community college. However, reality turned out to be a bit different from what I had imagined. You see, when you start college in the United States, some of the classes you took in high school count as part of your curriculum. Unfortunately, my high school credits from Brazil didn't transfer over, so I had to complete numerous general classes before I could focus on my engineering studies. This added two extra years to my college journey. Despite the challenges, I didn't give up. Eventually, almost 5 years later, I earned two associate degrees—one in engineering and the other in chemistry (Fig. 4).

Fig. 4 Author graduation photo; Cabrillo College Associate Degrees—Chemistry, Engineering

My time in community college was filled with a rollercoaster of emotions, uncertainties, and insecurities. English, being my second language, posed a significant hurdle. Learning everything took me much longer than it did for other students because I had to translate lessons, grasp new words, and understand unfamiliar subjects. Amidst it all, I also struggled with homesickness and felt a lack of support. There weren't many females in my classes, and I was often surrounded by male students. I distinctly remember looking around in classes and seeing only a few other female students. Moreover, I had some difficult experiences with a few male classmates. One of them once made a hurtful comment, suggesting that females might be slightly less smart than males. As you can imagine, this comment not only hurt my feelings but also made me self-conscious about speaking up because I was always afraid of saying something wrong. This experience made me feel like a minority in a different way, and I battled a lot of self-doubt throughout the process.

Being an adult in a foreign country presented its own set of challenges, but it also brought positive transformations. It was during my first semester at community college that I met my husband. We shared a class, and he took the bold step of asking me on a date. That marked the beginning of a beautiful journey, and 5 years later, we were happily married. He became my rock and my family—a constant support through life's challenges. In the eyes of my family, I might be seen as "Americanized," a testament to the growth and learning I experienced in this new culture. Despite the difficulties, community college turned out to be a place where I not only faced challenges but also found love and built a foundation for the life I have today.

College also opened a world of exploration and learning for me. It wasn't just about my engineering curriculum—I had the opportunity to delve into other subjects. I tried my hand at Italian for two semesters, immersed myself in a captivating microbiology class, and ventured into courses that weren't initially on my academic radar. It was a journey of discovery and, most importantly, a lesson in embracing failure as a stepping stone to success.

Facing my first academic setback was tough. Failing a class and having to repeat a semester felt like a major roadblock. There were moments when I questioned if engineering was the right path for me, and the thought of quitting crossed my mind. Thankfully, I had unwavering support from my husband, who reassured me that failure is not the end—it's an opportunity to learn and try again.

The second time around, armed with newfound resilience and determination, I approached the challenge with greater confidence. The experience not only prepared me better for future classes but also taught me a fundamental life lesson—failing is a part of the journey, and it's okay. Each difficulty I encountered strengthened me, and every failure became a valuable teacher. I learned that perfection isn't the goal; it's the mistakes that shape us and guide us toward continuous improvement.

4 Self-Discovery

After graduating, I found myself standing at a crossroads, questioning the path of chemical engineering and whether it still ignited my passion. I realized that the idea of spending most of my work hours in a lab didn't excite me as much as it once did. This realization opened a floodgate of thoughts—I discovered interests beyond the confines of the lab, leaving me in a state of confusion regarding my career path. Following a heart-to-heart conversation with my parents (as they were the ones financially supporting my education), I made the decision to take a year off after graduation, as my current student visa allowed me to do so.

During this year, my husband and I got married, marking a beautiful milestone in our lives. Despite my parents being in Brazil, they had supported me financially throughout college. However, it didn't feel right to lean on them when I wasn't pursuing further education. This realization fueled my determination to find creative solutions and make my own way in the world. This year off turned out to be not just a pause but a period of self-discovery, personal growth, and navigating the intricacies of adulthood.

I embarked on a journey of exploration, seeking out jobs with my associate in science degrees and exploring alternative paths. However, the job market surprised me, revealing limited opportunities for individuals with only associate degrees in science. To navigate through this period and contribute to our financial responsibilities, I took on various jobs, each serving as a piece of the puzzle to pay the bills.

After my year of exploration, a longing for science and the realization that a degree in engineering could open diverse career opportunities led me to make a pivotal decision. Uncertain of where this path would lead, I made the choice to transfer to San José State University (SJSU) and pursue my bachelor's degree in chemical engineering.

Starting at San José State University presented a new set of challenges. As a minority in many aspects, I again felt a sense of isolation and struggled to adapt. Once again, I found myself in classes with few females, and as the only Brazilian student, the feeling of being an outsider was prominent. The unfamiliarity with new terminologies and concepts heightened this sense of not belonging. The chemical engineering classes were more challenging than I had imagined, introducing me to subjects that were entirely new to my vocabulary.

One day, I connected with a fellow female classmate who, like me, was navigating the challenges of being a transfer student in pursuit of a chemical engineering degree. We both felt a sense of loneliness in our academic journey. During our conversations, she shared stories of a friend attending an engineering school in southern California and being part of a sorority called Alpha Omega Epsilon (A.O.E.), a social organization for women in engineering and technical sciences.

Undeterred by the absence of a local chapter, we decided to take matters into our own hands. We scheduled a meeting with the sorority organizers, who patiently explained the process of establishing a chapter. They outlined the steps, from becoming a group to eventually earning the status of a chapter. The journey involved fulfilling various requirements, such as creating laws for our sorority chapter, organizing professional events each semester, establishing committee chairs, and more.

Realizing that we had the opportunity to build something meaningful from scratch, we were motivated to embark on this journey. We saw it as a chance to not only shape a community in a way that resonated with us but also to extend support and sisterhood to other females who might be facing similar challenges. The decision to establish the Beta Upsilon Chapter of Alpha Omega Epsilon at San José State University was born out of a shared desire to create a space where women in STEM could thrive and find a sense of belonging.

Eager to make a positive impact, we embarked on a mission to connect with other like-minded women who shared our passion for STEM. I felt a strong sense of purpose, convinced that my experiences of isolation and judgment couldn't be unique. As a female in STEM, navigating through feelings of being underrepresented and contending with implicit biases that affected my confidence was challenging. We believed that there were more STEM females out there who might want the group we wanted.

With determination, we created vibrant flyers and strategically placed them throughout the science and engineering buildings. Additionally, we organized two meetings to share our vision and purpose. To our delight, the response exceeded our expectations, with over 15 female students attending our first meeting. This affirmed that our efforts were resonating with others who felt the same way.

Our initial meeting turned into an amazing discussion, bringing together females who shared our experiences and aspirations. During the gathering, we outlined the process of forming a sorority and the journey we would undertake before becoming an official chapter. Our criteria for welcoming new members were simple—a passionate student from a STEM major. We welcomed all interested females, creating a space where STEM major students could come together and thrive.

Taking on the role of president pushed me beyond my comfort zone. I had to speak up, lead meetings, and present ideas. I had to showcase the incredible potential of the sorority, enticing others to join our journey. The more I stepped out of my comfort zone, the more my confidence returned. It wasn't a seamless process, but the learning and personal growth were invaluable.

After putting in a lot of effort, our chapter, called the Beta Upsilon Chapter of Alpha Omega Epsilon Sorority, was finally started! We are the 44th chapter founded in the United States under Alpha Omega Epsilon. The chapters are named in Greek alphabet order, so after going through the alphabet, it becomes two Greek letters. That's how we became the Beta Upsilon Chapter. I was president of the sorority for 2 years.

Together, we focused on improving our soft skills and supporting each other. We organized different events for our group of sorority sisters, such as resume workshops, mock interviews, company tours, and networking activities. Our hard work paid off, and we became a popular sorority for women in engineering and technical sciences on campus (Fig. 5).

We also collaborated with other student organizations in the engineering building and connected with companies for tours, mock interviews, and more. Exciting opportunities just kept coming our way!

One of the most amazing experiences I had during my time in the sorority was the chance to share my story at a Women in Engineering Conference event at San José State University. This annual event, organized by SJSU, brings together companies from all over the Bay Area with the goal of connecting female engineering

Fig. 5 The day that our Chapter was founded, Beta Upsilon Chapter of Alpha Omega Epsilon

students with potential employers. Belle Wei, the organizer of the SJSU Women in Engineering Conference, reached out to us and invited us to participate.

During that particular year, over 400 people attended, including students, faculty, and employees from various companies. I had the honor of delivering my speech just before the final dinner speaker to the entire audience in the ballroom. Even today, the current president of our sorority continues to speak at this event annually.

Being part of the sorority exposed me to companies and diverse career options. We organized and attended numerous events in the Bay Area to help shape our future careers. This exposure opened doors to different career paths and introduced me to a multitude of people. I attended various networking events and conferences, meeting professionals with unique career paths. These events not only bolstered my resume but also enhanced my interview skills through mandatory mock interview and resume review sessions.

Surprisingly, being involved in starting and running a sorority exposed me to the world of sales. It felt like launching a sales venture, where we sought support from the university and staff for sponsorship. The entire process involved a year of meticulous planning and creating a roadmap for the sorority until its official founding. This unique experience helped me understand myself better. In a technical sales position, you are not alone; you need the support of various roles like marketing, application scientists, field engineers, product specialists, and more. Collaboration with different people is essential to ensure you can meet your customers' needs and offer them the best solutions for their requirements. The success of our sorority was deeply rooted in the partnerships we formed with school staff, the sorority's main board, and other student organizations.

5 Evolution

In my quest to find a career that resonated with me, I decided to explore beyond my major. The lab environment no longer provided the complete fulfillment I sought, so I ventured into new territories. Sales engineering, or technical sales, caught my attention. Given my enjoyment of interacting with people, my outgoing nature, and my competitive spirit, the prospect of working toward and achieving sales quotas excited me. To explore into this field, I opted to pursue a minor in business alongside my chemical engineering degree at SJSU (Fig. 6).

The minor allowed me to tap into different facets of my abilities and embrace a novel experience. With the flexibility to select five classes for the business minor, I chose courses that could help me navigate the world of sales. The business classes, especially Fundamentals of Management and Organizational Behavior, broadened my perspective and offered insights into human behaviors. I distinctly recall my initial confusion when faced with a homework assignment that stated, "there is no right or wrong answer." I approached the teacher, expressing my belief that there's always a correct answer. She explained that unlike in my engineering assignments where equations yield a single answer, when dealing with people, there can be

Fig. 6 Graduation Day, San Jose State University; Bachelor's degree in Chemical Engineering with minor in Business

multiple possible outcomes. This realization was crucial in shaping my outlook on life, emphasizing that, unlike in math, there aren't always clear-cut right or wrong answers. Acknowledging the existence of different answers and paths fostered an open-minded approach to my career choices.

As I contemplated a career in sales, I also enrolled in a negotiation class, which turned out to be immensely enjoyable. The class incorporated numerous negotiation role-plays, and I found them to be both insightful and entertaining. Balancing these diverse classes with my regular chemical engineering studies was both interesting and fulfilling.

At San José State University, the chemical engineering program offers several specialized tracks, including semiconductor, environmental, energy, and others. I opted for the bioengineering and biochemical track because the classes it offered seemed more captivating than those in other tracks. Little did I realize that this decision would lead to my first exposure to the fascinating world of biopharmaceuticals (biopharma).

While juggling my classes, I made it a point to actively participate in extracurricular activities. Attending conferences, company tours, and workshops became integral to my educational journey. Given our sorority's emphasis on professionalism, we incorporated specific requirements into our bylaws, mandating attendance at career development opportunities. Annually, I ensured attendance at a minimum of one conference, with the San José State University Women in Engineering Conference being nonnegotiable due to its on-campus convenience and our special invitation from the organizers. Additionally, I participated in 3–5 company tours each year. Bay Area companies often extended invitations for these tours, either by visiting the engineering and science buildings or reaching out to student clubs like ours.

Collaborating with other student organizations, our sorority organized workshops featuring at least one mock interview and one resume session per semester. To enrich these events, we invited alumni and professionals from various fields to serve as panelists, sharing their stories, insights, and tips. These opportunities proved invaluable, offering me a chance to enhance my personal development and gain insights into diverse career paths.

One standout experience occurred at the San José State University Women in Engineering Conference, where I was particularly inspired by a speaker named Dr. Darlene Solomon. Typically, conference speakers at SJSU lean toward software engineering, but Darlene stood out by focusing on life sciences and biology—subjects that resonated deeply with my interests. At the time, she served as the Chief Technology Officer (CTO) of Agilent and passionately discussed synthetic biology's significant role in addressing global challenges. Synthetic biology involves engineering organisms for various purposes and finds applications in diverse markets. This concept fascinated me, especially considering my enjoyment of bioengineering classes and the growing passion for the biological aspects of my major.

Darlene's humbleness, intelligence, and approachability left a lasting impression and prompted me to delve deeper into her story and the company she represented, Agilent. Subsequently, our sorority had the privilege of visiting Agilent's Santa Clara site for an enlightening company tour. The Agilent tour provided a unique opportunity for me to meet employees who genuinely loved their work. The company's rich history of innovation, coupled with the passion and happiness expressed by its employees, made it seem like the ideal place for me to thrive.

6 Destiny

In that same year, I decided to extend my senior year into 2 years to accommodate a part-time internship. At the time, I was juggling various jobs unrelated to my major. Living just 10 minutes from San José State University with my husband, who had already graduated and was financially stable, I had the flexibility to seek an internship, even if the pay wasn't substantial. I was determined to gain more experience and explore the field of sales. Internships, as I soon discovered, are incredible

Fig. 7 Author with Dr. Darlene Solomon, Agilent CTO and Senior Vice President (retired July 2023)

opportunities to uncover what we do or don't want to pursue postgraduation. A year later, at the annual Women in Engineering conference, Darlene Solomon (Fig. 7) returned as the keynote speaker, and my excitement was palpable. I wanted to connect with her and glean insights into her remarkable career path.

Although my search for her at the conference proved unsuccessful, destiny had a surprise in store. Upon arrival, attendees were assigned numbers for seating during lunch and dinner. Fortunately, my number coincided with those of several Agilent employees, landing me at a table with them. Engaging in conversation, I shared my aspirations for a career in technical sales. Encouraged by these Agilent professionals, I decided to apply for an intern inside sales position at Agilent. With their recommendation, I secured the position, which involved supporting the chromatography sales team in the United States.

Throughout the internship's one-year duration, I immersed myself in the intricacies of Agilent's instruments—learning how to configure them, understanding their differences, and grasping their applications across diverse markets. I gained insights into prospecting new customers and creating quotes. This internship, which extended beyond the typical summer stint, turned out to be the perfect fit. It allowed for full-time engagement during the summer and part-time commitment during the school year, marking the true commencement of my journey into the field of sales.

Starting my first job at a worldwide company known as the leader in life sciences, diagnostics, and applied markets was incredibly intimidating. Amidst the vast ocean of Agilent's extensive workforce, boasting over 18,000 employees and covering over 25 countries, I initially felt like a small fish. However, as the days and months unfolded, I discovered that there was no need for anxiety.

My team, despite the company's enormity, consistently proved to be approachable, always willing to address my questions and attentively listen to my input. Colleagues were exceptionally helpful whenever I sought guidance, and my manager provided constructive feedback, fostering my professional growth. I committed myself to absorbing every bit of knowledge, diligently working toward success in every task that came my way. Recognizing this as a unique, once-in-a-lifetime opportunity, I was determined to make the most of it.

The ensuing year unfolded as a whirlwind of activity, with classes, internship commitments, and sorority engagements dominating my schedule. Despite the demanding months, I found contentment in the abundance of learning and networking opportunities Agilent offered. Actively reaching out to fellow employees through emails, I discovered numerous positions within the company. The wealth of possibilities and avenues for personal and professional growth excited and motivated me to strive even harder.

This period of exploration revealed the flexibility and adaptability within Agilent's career paths. The realization that I could leverage my engineering degree for success in sales while leaving room for potential shifts to areas like business development intrigued me. The company culture echoed my initial observations during my student visit, where employees appeared genuinely content, maintaining a healthy work-life balance. Agilent provided a space where technical skills could thrive, and the well-being of employees was prioritized, a far cry from being treated as mere numbers. People seemed very open to exploring and trying new career paths; our current CEO, for example, started in sales.

7 A Career Path

Four months before my graduation, the quest for a full-time job at Agilent commenced. My manager, Lara (Dubuc) Stephens (Fig. 8), played a pivotal role in supporting my aspirations. Lara provided constructive feedback during our meetings, propelling my personal and professional development. Through hard work and dedication, such as achieving the highest number of quotes and customer calls among interns, my efforts garnered attention.

The big moment happened when Lara told me about a brand-new job—a Biopharma Market Specialist role. This special role meant I would be an inside sales representative, connecting with biopharma companies in California. It matched

Fig. 8 Author with Lara Stephens, CLSD Global Commercialization Manager—Agilent

exactly what I had dreamed of doing. It meant I could try out different things and make an impact, even though I was still pretty new to the company.

When Lara explained that this role involved selling to biopharma companies in California, I couldn't help but get excited. It was like hitting the jackpot—a job that mixed sales, science, and the chance to explore new ideas, all from the beginning. It gave me the freedom to experiment and prove that I could make a difference, even though I was a newcomer. And just like that, my official journey into the exciting world of the biopharma industry began.

Working in sales has been an incredible experience for me as it allows me to combine my technical skills with communication and people skills. However, when I started, I soon realized that selling to scientists in the biopharma field was different from selling regular products. For example, working in a clothing store, you have the customer try on an outfit and tell them if it looks good or not. The sale of an instrument that can be configured to help customers develop new drugs and vaccines is much more complex. It involves engaging in conversations with application scientists, product specialists, and others to help customers find the right solutions to their problems.

When I started this job, the world of biopharma was like a mystery to me. So, I decided to become a bit of a detective and learn as much as I could. I watched training videos in Agilent's database and talked to many experts at Agilent. I adopted a mindset of continuous learning. I realized that to be really good at my job, I needed to know a lot about Biopharma.

Learning about the biopharma market gives me hope for a better future, with the possibility of finding cures for many diseases that still impact our society, like cancer. It's amazing to see how biopharma researchers are continuously exploring new ways to tackle these challenges.

When the COVID-19 pandemic hit, many scientists and companies in the biopharma industry were working tirelessly to develop a vaccine. It was a tough time for our society, but talking to these dedicated scientists gave me hope and made me realize the importance of their work in saving lives. I had lots of calls with customers who were using our systems to work on the COVID-19 vaccine and tests. It makes me proud to say that many companies with solutions for COVID-19 use Agilent instruments.

Looking back, I'm glad I chose a career path in biopharma. The best part about it is that there's still so much more to be researched and developed. All in all, biopharma is an exciting and promising field that keeps pushing the boundaries of science and has the potential to create a better, healthier world for all of us.

As my career began to take off, I found myself missing the support system I had during my college days, like the sorority. Fortunately, at Agilent, we have Employee Networking Groups (ENGs) that provide me with a similar support system in my professional life. As a site employee, I had the wonderful opportunity to come across engaging flyers for various events organized by ENGs, prompting my decision to join. Having a support system, especially as a minority, is crucial, and it helps me stay confident as I face new challenges and opportunities. I believe in the importance of continuous learning and adaptability, so I decided to join the Women@Agilent ENG. This group has given me many opportunities to network with people from various departments and different countries. I had this amazing opportunity to meet my idol, Darlene Solomon, and interview her during an event called "I Wish I Had Known," organized by the Women@Agilent ENG. It was a virtual event where I got to ask Darlene questions about her career path and learn some tips and tricks—things she wished she had known when she started. Being able to freely ask her about career growth and hear her advice was such a fantastic experience. Through the Women@Agilent ENG, I've learned valuable lessons about business acumen, empowerment, leadership, and more. For example, one big lesson is about business acumen, which is like having a special set of skills for understanding how a business works and how you fit into it.

8 Piecing Together a Puzzle

Imagine working in a big store. It's not just about doing your job; it's about understanding how the whole store operates. Similarly, in a company, it's crucial to know not only what you bring to the table but also how the entire business functions. Learning about business acumen has been like discovering the secret sauce to success. It's about knowing the different parts of the company, almost like puzzle pieces fitting together. This understanding not only helps me do my job well but also allows me to connect with people in other departments. It's like having a backstage pass to the whole show.

For instance, I realized I needed to team up with other salespeople at Agilent who sold different instruments but were still part of Agilent. This connection opened new doors for me. They introduced me to new customers, and I learned more about what Agilent offers beyond my own department. It's like building a network of friends who all help each other out. In a nutshell, business acumen is like having a superpower. It lets me support my customers better, and at the same time, I get to represent my company in the best way possible. It's not just about doing a job; it's about understanding the whole picture and making the whole team shine. Just like the sorority A.O.E. during my college days, this ENG has been a significant source of growth and support in my career.

I am also a part of La Chispa, an ENG for Hispanics and Latinos, and that has been an amazing experience for me. It's like finding a community that understands and shares my cultural background. In La Chispa, I've connected with other Brazilians who get what it's like to feel isolated, homesick, experience culture shock, and struggle with not quite fitting in. But it's not just about Brazilians; I've also had the chance to learn more about my fellow Latinos and Latinas. It's surprising how we can all have so much in common while also being different in our own unique ways.

For my foreign Latina side, La Chispa is like a supportive home, a place where I feel understood. On the other hand, being a part of Women@Agilent allows me to look up to other females, helping me feel like I'm not alone in my journey. These networking groups have turned into fantastic extracurricular activities for me. They help me grow both personally and professionally, build my network, and learn from others.

More than anything, these groups make me feel welcomed and at home. In this journey to become a better version of myself, I never feel alone, thanks to the sense of belonging and support from La Chispa and Women@Agilent.

9 Career Moves

After a few years in my first full-time position, I had the opportunity to apply for a position as a biopharma account manager, which is an outside sales role. Inside sales reps work from a distance with the customer and are usually located at the office or they work from home. They use digital tools to connect with customers, while outside sales reps are out in the field, meeting customers face-to-face and building strong, personal relationships. As an inside salesperson, I supported different account managers, while in this position, I have the chance to own and manage my territory and accounts. I have someone on the inside sales team supporting me while I am on the road. It's important for me to maintain and strengthen relationships with existing customers. Today, I am the main point of contact between my customers and Agilent, so I need to ensure that their needs are met and that they are satisfied with our instruments. It's been an eye-opening experience as I get to learn more about how my customers work and understand their specific needs. I particularly enjoy visiting laboratories and seeing my customers in action (Fig. 9). I collaborate closely with numerous colleagues to help our customers find the perfect solutions for their needs while witnessing their growth and involvement in amazing projects. For example, depending on my customer's needs, I have specific application scientists and product specialists who can help me. Those colleagues are focused on specific products or areas, allowing them to have deeper conversations

Fig. 9 Author, Biopharma Account Manager—Agilent

with my customers so that together we can find the best solution to meet the needs of the customer (Fig. 10).

One aspect of my job that I truly cherish is supporting start-up companies (smaller companies with less than 100 employees) and being involved with incubators and accelerators. It's rewarding to see these ventures take off and contribute to the field of biopharma in innovative ways.

Incubators and accelerators are special places that offer valuable support and assistance to early stage companies. When companies join incubators or accelerators, they usually pay a monthly fee. The specific perks and benefits they receive depend on the program and location they choose. Incubators, for example, are where many small companies start when they don't have a business model yet. These incubators provide companies with a physical space, and sometimes they even offer a communal lab space. If you're starting a business and don't have much funding, you can rent a space in an incubator and use their communal lab with instruments. This can be a significant cost saving, as lab equipment can be quite expensive. This is especially helpful for start-ups that need to focus their limited resources on their research and development.

On the other hand, accelerators, as the name suggests, are meant to speed up the growth of companies. They usually work with small companies that are a bit more

Fig. 10 One of the many events that Agilent participated to meet new customers, photo shows the author and her coworkers at the table

established. Accelerators are there to help these small companies expand. They can go a step further and provide intensive support to the companies they work with. This support can sometimes include mentorship from experienced professionals, guidance on business development, assistance with regulatory processes, and access to a vast network of industry partners and potential investors.

Different accelerators may have various rules and offerings. For instance, some accelerators only allow companies to stay for a specific time (for example, up to 18 months), and each year, they may choose one company to receive a substantial cash prize, providing additional funding. The main goal of both incubators and accelerators is to help these companies grow and succeed by providing them with a supportive environment and essential resources. Many of the individuals who own these spaces often become investors and board members of these companies, potentially bringing them financial success in the future. These spaces are particularly beneficial for biopharmaceutical companies with innovative ideas aiming to bring new drugs and therapies to the market.

In the biopharmaceutical industry, both incubators and accelerators play vital roles in fostering innovation and entrepreneurship. They create an environment where new ideas can flourish, and young companies can receive the help they need to grow and succeed. This, in turn, contributes to advancements in healthcare and the development of new medical treatments, which can benefit countless people around the world. A notable example of a company that began in an incubator is Moderna, a company that had a significant impact during COVID-19 by developing a vaccine to assist the world in addressing the pandemic [1].

In the future, I want to keep working in the biopharma market, particularly with incubators and accelerators. These places provide essential support to early stage companies, helping them develop their ideas and create new drugs and therapies. Right now, there isn't a specific position focused on working with incubators and start-ups at Agilent, but I'm determined to think outside the box and create new opportunities.

10 I've Reached My Destination

After 15 years in the United States, I can confidently say that I've found my home. I'm happily married to my amazing husband, Johnny, and we have Penelope, our dog daughter (Figs. 11 and 12). Looking back, I've learned a few things that I'd like to share as advice.

First, don't let fear stop you from pursuing your dreams. The world can be scary, but doing things that scare you can bring you the biggest rewards. Moving to the

Fig. 11 Isadora with her husband, Johnny, and their dog Penelope

Fig. 12 Author, author's parents, and her in-laws in Rio de Janeiro, Brazil

United States for college was terrifying, but it led to significant personal growth and made me a stronger woman.

Second, don't be afraid to ask for what you want, even if it seems simple. Always remember if you don't ask, the answer is already no, so why not ask and maybe receive a yes? I would never have imagined that Darlene Solomon would talk to me or would say yes to attend an event for the A.O.E. sorority as a speaker. You never know what opportunities might come your way.

Lastly, trust that your time is your own, and you don't need to compare yourself to others. Don't try to be on anyone else's schedule. When a butterfly is emerging from its cocoon, she needs to get out of it in her own time. If you try to help her out of the cocoon, you will hurt her. We all have our own time to grow and succeed. It took me longer than expected to finish my degree, but in the end, it was the best decision. It helped me discover my true career passion, and I graduated with confidence and secured a job that I truly love.

Reference

1. Primack, D., & Herman, B. (2020, March 17). *The company leading the race to a coronavirus vaccine*. www.axios.com/2020/03/17/moderna-coronavirus-vaccine-trial

Breaking into Sales and Biopharma: An Early Career Woman's Journey

Sharon Basil

Contents

1 Introduction

The biopharmaceutical (biopharma) industry is a dynamic and rapidly evolving field that plays a pivotal role in advancing healthcare and improving the lives of individuals worldwide. This chapter will explore the experiences and journey of the author as a woman in biopharma, sharing her experience in the industry—specifically in sales. Like many sectors, the sales industry has historically been dominated by men. This presents unique challenges and obstacles for women. The author will describe her experiences in the sales and biopharma industry while highlighting her successes and accomplishments.

Despite significant progress made in recent years toward gender equality, women still face persistent barriers and unconscious biases in their professional pursuits. Within the sales and biopharma sector, these challenges can manifest in various forms, including limited access to career advancement opportunities, lack of

S. Basil (✉)
Agilent Technologies, San Francisco, CA, USA
e-mail: Sharon.basil@agilent.com

S. Taylor, M. Campbell (eds.), *Women in Biopharma*, Women in Engineering and Science, https://doi.org/10.1007/978-3-031-80348-2_7

representation in leadership positions, and subtle or overt gender biases that affect workplace dynamics. The author has personally seen and been a part of sales teams with few or no women, management structures dominated by men, and has experienced different expectations imposed on women versus men.

Against this backdrop, this chapter presents a personal narrative of the author's career journey in the biopharma industry. It provides insights into the educational background, career choices, and experiences she went through to overcome obstacles encountered along the way. She also reflects on her past self and considers feedback and advice she wished she had received when she was younger, and she shares her advice for readers. By sharing these experiences, the chapter aims to shed light on the unique challenges faced by women professionals in the biopharma and sales industry and highlight their resilience and accomplishments.

The author would like the reader to keep in mind that her journey is through the eyes of a professional who still considers herself to be early in her career (graduated from college 3 years ago). While this journey is not coming from one with many years of experience in the industry, she hopes to provide an alternative perspective and relate to others who are also beginning their careers and figuring themselves and their careers out.

2 Education

The author began her career journey when she attended the University of Delaware from 2016 to 2020. When deciding on her major, she knew she wanted to influence health and improve the lives of the general community while also utilizing an engineering background. The University of Delaware provided a major that she saw fit her interests with biomedical engineering.

Upon entering her first biomedical engineering course (BMEG101), she was excited to see the gender population was pretty evenly split, a rare occurrence within engineering majors. The cohort was small, about 30 students. Throughout the 4 years, she formed strong bonds within the group that she continues to this day, as well as felt inspired and empowered by professors and mentors she was taught by along the way.

The highlight of her time as a biomedical engineering student was her senior design project, where she and a group of three other students were tasked with creating an MRI (Magnetic Resonance Imaging) compatible Oscillatory Lower Body Negative Pressure Device (OLBNP) for the Neurovascular Aging Lab at the University of Delaware. An OLBNP is a negative pressure vacuum chamber surrounding a subject's lower body, and its purpose is to use fluctuations in negative pressure to alter blood flow to the brain. The device built was a successful and functional product and became part of the lab's grant to help sponsor and conduct future research (Fig. 1). Ultimately, the device aided in developing interventions for delaying or reversing age-associated cognitive decline (Fig. 2).

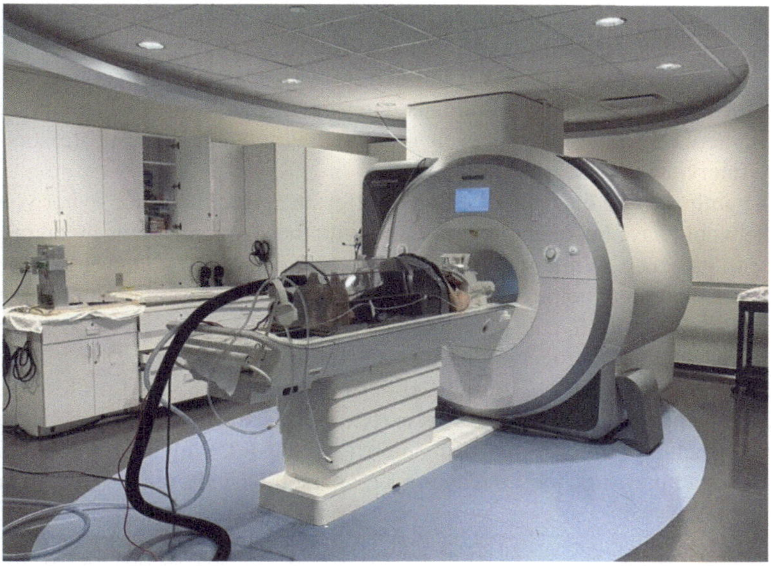

Fig. 1 OLBNP Device built by the author and senior design team

Fig. 2 OLBNP device built by author and senior design team

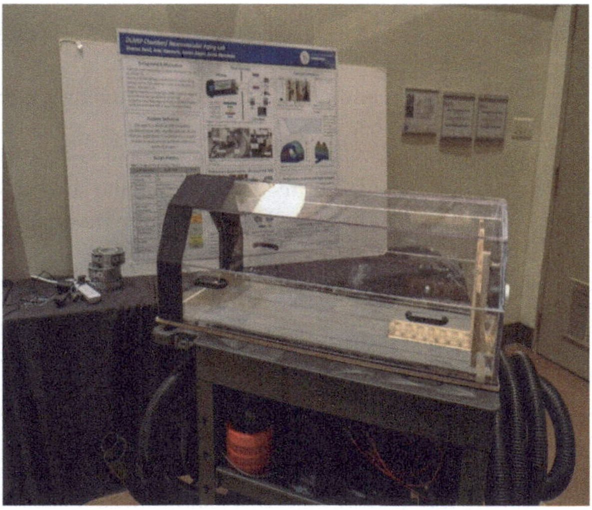

This senior design project was a special experience for the author, as it was one of her first hands-on experiences working in a machine shop to build something large-scale with a lasting impact. The experience felt empowering, as many of her mechanical engineering counterparts were male. Seeing that she could also successfully build an instrument under challenging circumstances and a limited budget was one of the first experiences where she felt empowered.

In addition to courses, she loved that the university provided valuable female student organizations, such as the Society of Women Engineers (SWE) and AOE (a female engineering sorority). SWE and AOE provided valuable career development events for female engineering students, such as meeting potential employers on campus, networking events between other female students and professors, and resume/interview prep sessions.

If the author could change one thing or offer one piece of advice to her younger self, she would have become more involved with these organizations by taking on a leadership presence. While she got a lot from participating in the events presented by the organizations, looking back, she wishes she had spent more time building a community with other female engineering students with similar passions and being part of the team that put together these events for the student body. Doing this could have expanded her network, introduced her to more students and professors, and helped her gain more leadership skills early on. Realistically, she knows not to be too hard on herself for not participating more than she did. With her rigorous coursework and other part-time jobs, the author kept herself busy during her time in undergrad in many ways.

3 Early Work History

The author always had a strong desire to join the workforce as soon as possible. Her first job was in high school as a tutor to 3–5-year-old children at Kumon, where she taught them basic reading and math skills. Her future roles as a sales lead in retail, lab technician, and background in biomedical engineering led her to find her passion in biotech sales. She learned that she had enthusiasm for serving others, skills in communication and sales techniques, and the academic knowledge needed to best serve clients in the biopharma industry. Upon searching for her next summer internship, the summer before her senior year of college, she found that Agilent was offering an inside sales internship on the Chromatography and Mass Spectrometry Instrumentation team.

The author was excited to join such a well-respected, global biotechnology (biotech) company like Agilent. Agilent is a global scientific company headquartered in California that specializes in providing analytical instruments, software, services, and consumables for laboratories, enabling advancements in life sciences, diagnostics, and applied chemical markets. One of her career goals was to serve healthcare and improve the lives of others. As an inside sales intern, she worked directly with Agilent's customers and felt that she was making a direct impact. Her duties included assisting customers over the phone to qualify new instrument purchases, helping account managers create quotes, and prospecting for new instrument opportunities.

Her internship at Agilent's Delaware Little Falls Site gave her an incredible opportunity to meet individuals and leaders in the industry from diverse backgrounds, learn about their experiences and their career journeys as well. There were many networking events, career development panels, mentors, and presentations for the interns that enhanced their experience as an intern. One of the ones that stood out to her was a panel with previous interns describing their internship at Agilent and where they are today at the company. It was inspiring to see the different career opportunities at the company and how far individuals who started out as interns have taken it. We were also given the opportunity to meet our CEO, Mike McMullen, at the company's Christmas Party in 2019. It was humbling to know that many people go their whole careers without having the opportunity to interact with the CEO of their company, but our intern team was able to meet him in such a casual setting.

4 Agilent Journey

As the author's internship was winding down and she was near graduating college, she needed to find a full-time job. The author had a strong desire to work full time on the inside sales team and there were great opportunities at the time. After her internship, she realized she had a passion for technical sales, and was excited about the opportunity to work at Agilent full time on the sales team. The author had a journey with ups and downs where she got turned down from a couple of positions due to the competitive nature of seeking a job during peak college graduation season. Additionally, graduating in 2020 during Covid-19 provided additional challenges, including hiring freezes due to budget cuts, and being in a more competitive interview environment with more experienced individuals who had lost their jobs. But she finally landed her dream job as an Inside Sales Account Manager after a few months of interviews.

Looking back on her experiences of getting turned down from multiple jobs from other teams, she was actually thankful for those experiences. Those experiences taught her that nothing in life comes easy or is guaranteed. As a woman of color raised by immigrant parents, she did not receive the typical encouragement to be proud and confident in her accomplishments as others might have. She was taught to be hardworking and humble about her achievements.

However, through her experiences, she learned that when searching for your next step, it is critical to give it your all and proudly showcase your accomplishments instead of shying away from them. In future interviews and performance reviews, she learned to effectively communicate achievements by highlighting specific experiences and quantifiable results and practicing presenting ahead of time. She was thankful for the mentors at Agilent she had throughout her internship who gave her the confidence and inspiration to be proud of her achievements. These mentors included individuals on her direct inside sales team, and others she had networked with throughout her internship. Because of them, she was able to perform strongly in her future interviews and secure a full-time role as an Instrumentation Inside Sales Account Manager supporting the Southern California district. This role is a common next step after the Inside Sales Internship and provides the inside sales account manager with the responsibilities of owning a territory, having a quota, and owning sales from start to finish.

The author looks back on these events with great satisfaction, as they made her a stronger and more confident individual. Because she didn't receive some of the roles she applied to in Delaware, she received a unique and once-in-a-lifetime opportunity to move to California for her role as the Inside Sales Account Manager. She never thought moving across the country was something she was brave enough to do, but after living at home during the pandemic for so long, she was excited to start a new life after college in a new place. Luckily, she also had a few friends from college who had moved to the Bay Area and had support from coworkers on the Santa Clara site. After all of her personal growth and experiences, she took on this challenge and made it well worth it. While she missed her family and friends, she took this opportunity to explore new places, meet new people, and become more independent. Overall, she is grateful for all the events that led to where she is now.

5 Women in Sales

After her experience in sales, the author wanted to discover if her experience of being in a male-dominated industry was unique or universal. Some quick research shows that there are many other females who might have faced similar challenges to the author. Below, Fig. 3 outlines the gender demographics of sales roles for all industries, from individual contributors to executives.

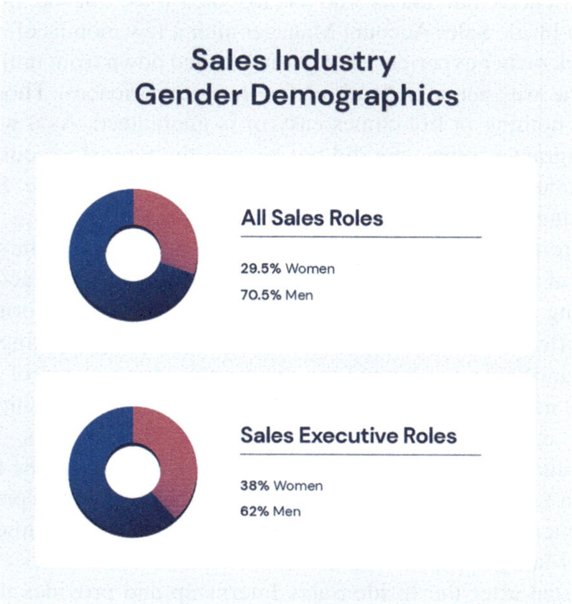

Fig. 3 Graphic of demographic of sales industry demographics [1]

Furthermore, although women generally outperform and achieve higher quotas than men, and also earn slightly higher commission rates, they're typically paid less overall compensation. This information was pulled in February 2023 from Zippia [2]. To summarize additional information from this source:

- Male-led sales teams are 76% men and 24% women versus the general population 50/50 split.
- Female-led sales teams more closely reflect the general population at 48% women and 52% men.
- Women achieve 8% higher quota attainment than men.
- 78% of men regularly achieve their sales quotas, while 86% of women achieve quota.
- Women in sales outperform men by 3% but are paid 7% less.

Being a woman in the sales industry brings both triumphs and challenges. While women tend to outperform men in achieving higher quotas, they often face disparities in compensation. The composition of male-led sales teams is predominantly male, while female-led teams come closer to reflecting the general population split. Despite these imbalances, women consistently demonstrate their capabilities by achieving higher quota attainment compared to their male counterparts.

These statistics shed light on the resilience and accomplishments of women in sales, showcasing their ability to excel in a challenging and competitive environment. However, it is disheartening to note that despite their performance, women face compensation disparities. While women outperform men, they earn 7% less overall. This discrepancy is particularly evident in the base salary. These findings underscore the importance of addressing gender-based pay gaps and ensuring equal opportunities for women to receive fair and equitable compensation for their contributions in the sales industry.

When examining the author's experiences as a woman in the sales industry within the biopharma sector, she too encountered some of these observations. Although it is unsure how compensation differs between men and women at the company due to other factors like job experience being the leading decision maker of salary, it was clear that there were significantly less women in managerial positions, especially on sales teams. However, she can speak to her own success in her role by achieving her sales goals and contributing to the success of her district. Her experiences serve as a testament to the capabilities of women in sales, while also emphasizing the need for organizations to address pathways for more women to move up the career ladder and enter more managerial roles.

6 Women in Biopharma

There is hope! A 2022 research study indicated that in biopharma companies, 46% of employees are women. However, the research also shows that this proportion shrinks to just 6% at the CxO level. These statistics indicate that, while men and women educated in the biopharma industry enter the workforce at the same time, women's careers plateau quicker than men's [3]. This may be attributed to systemic challenges such as gender bias and the struggle to balance work and family responsibilities, emphasizing the importance of addressing these issues to foster greater gender equality in leadership positions.

It is not surprising to read this, and it is definitely not specific to the biopharma industry. The author believes that we are making progress overall toward bridging that gap with the formation of organizations dedicated to helping women in STEM (Science, Technology, Engineering, and Mathematics) network, develop their careers, become leaders, and gain confidence. Some of these organizations include the Society of Women Engineers (SWE), the Association for Women in Science (AWS), and many more.

One of the most prominent organizations that she was a part of in undergrad is SWE, which has an annual national conference where women in STEM can network and learn skills to develop their careers. Agilent proudly sponsors a SWE Employee Network Group (ENG) within the company, sends a delegation of women to the national SWE conference for personal and professional development, sponsors a booth at the conference career fair, and supports SWE outreach activities designed to encourage middle and high school-age girls to consider careers in STEM fields. In addition to the SWE ENG, Agilent has other employee networking groups, including the Women@Agilent ENG, where the author is a core team member for the Santa Clara Site Chapter. This group provides networking and mentorship opportunities internally for employees, as well as externally for the local community. Being a part of these networking groups has given me opportunities to meet other females in my company, share advice and experiences, and be a leader in my community.

Although there is much work to be done in closing the gender gap at the executive leadership level in biopharma companies, the work being done through women around the country to slowly push the needle is definitely making an impact and creating a new generation of women leaders.

7 Conclusion

The experiences the author shared in this chapter allowed her to reflect on her personal education and career journey and how she navigated becoming a sales professional in the biopharma industry. Through her personal narrative and insights, she explored gender biases and obstacles that women are often faced with throughout their careers.

Despite these challenges, the experiences and achievements of women professionals showcase their resilience and commitment to serving and advancing innovation in the industry. By actively engaging in networking and mentorship initiatives presented to them throughout their careers, we are slowly paving the way toward success for future generations.

The author imparts valuable advice for women in STEM and sales based on her experiences. First and foremost, she emphasizes the importance of confidently showcasing achievements and contributions during interviews and performance reviews, underscoring the significance of preparation and self-promotion in professional settings. Furthermore, the author highlights the value of resilience in the face of setbacks, sharing her own experiences of missed job opportunities and expressing the belief that, in the end, everything tends to work out as it should. Lastly, she encourages women not to be disheartened by gender gap statistics but rather to focus on giving their best and actively engaging with organizations working toward workplace gender equality.

While these experiences and reflections are completely personal and, of course, not indicative of everyone's journeys, she hopes that it sheds light on being a woman in STEM and biopharma, explores some of the challenges faced, growth opportunities, and provides inspiration to others in similar situations.

References

1. Patel, S. (2023). *Women in sales: How female leaders are empowering the sales industry.* Mailshake. https://mailshake.com/blog/women-in-sales/
2. Kolmar, C. (2023, February 26). *100 Important sales statistics, figures, and salaries.* https://www.zippia.com/advice/sales-statistics/#:~:text=Sales%20Statistics%20by%20Gender&text=Female%2Dled%20sales%20teams%20more,but%20are%20paid%207%25%20less
3. Frontline Genomics. (2023, July 5). *The lack of gender equality in biopharma.* Frontline Genomics. https://frontlinegenomics.com/the-lack-of-gender-equality-in-biopharma/

The Biopharma Industry Around the World

The Biopharma Industry in India

Veenaa Jayaraman

Contents

1 My Odyssey into Indian Biopharma

"Phew! What's that smell?" I wrinkled my nose and turned to my colleagues. This was my first visit to a large biopharmaceutical manufacturing company site in Bangalore. The year was 2010, and I had moved to Bangalore in my new role as the Senior Marketing Executive for GE Healthcare Life Sciences. I loved tagging along with the sales folks on customer site visits and interacting with them directly to understand how my marketing communication messages and campaigns are perceived by them. "Haven't you ever smelt *Escherichia coli* (*E. Coli*) before?," asked my colleague, while smiling and pointing to the left. My jaw grew slack as I stared at massive tanks, lined up like sentinels, towering over the neighboring corporate buildings. I learned they were bioreactors with a capacity of 50,000 liters, growing

V. Jayaraman (✉)
Bioresearch, Spinco Biotech, Chennai, Tamil Nadu, India

GE Healthcare Life Sciences, GE, Bangalore, India

Lifesciences, Beckman Coulter, Mumbai, India
e-mail: veenaa.jayaraman@spincotech.com

S. Taylor, M. Campbell (eds.), *Women in Biopharma*, Women in Engineering and Science, https://doi.org/10.1007/978-3-031-80348-2_8

bacterial strains of *Escherichia coli* to produce molecules of insulin and its analogues. Until that moment, my life had been about test tubes, culture plates, and flasks with DNA or proteins run on gels, invisible to the naked eye. The fact that the science experiments that I had carried out in the lab could lead to to this scale of production of a life-saving drug was truly a revelation. No amount of reading and theory had prepared me for this. I felt newly energized, and fortunately, I still feel the same energy years later when I see biotech drugs come to life (or market). From then on, to quench my thirst for more exposure to such real-world applications, I visited many such biopharma production sites in other cities across the span of India and immersed myself in learning about manufacturing on a large scale. I explored vaccines manufacturing, microbial and cell culture monoclonal antibodies production etc. The concept of how a modified microbe or cell in a laboratory environment can be magnified at an exponential scale, without losing the key quality of the recombinant protein it produces, became my life's work. Scalability and quality at scale became my passion in this industry.

I was fortunate to be at the right place and in the right company, as it allowed me to meet different genres of people—academic researchers from universities, research and development scientists and manufacturing professionals from the industry. With GE's technology offerings fir biopharmaceuticals, fitting right into my passion, I was able to offer solutions to the challenges faced by our customers while taking a biological drug to market from scratch. Bringing the latest technology home to enable my country's scientists has been a rewarding life experience, all starting with that innocuous incident of smelling *E. coli* from the tanks.

Biotechnology came into my life when I was a 10-year-old girl. My parents were both academics and biochemists by profession. They met during their PhD tenures at the famous Indian Institute of Science in Bangalore—two completely contrasting, yet extraordinary minds. My father, Dr. J. Jayaraman, was the perfect professor, with a love for teaching and with an ability to explain biochemistry in the simplest of manners. He was also the perfect gentleman, never raising his voice or losing his temper with his students or staff. Instead, he was the go-to person for everyone to rant about their woes or get advice about a problem. With multiple publications and a very popular book to his name, he soon became the Head of the Department of Biochemistry at Madurai Kamaraj University, a fast-growing educational institution in the southern part of India. My mother, on the other hand, was a fiery, 'fast-n-furious' lady, who was equally brilliant and dynamic (Fig. 1). My parents were fortunate enough to get their tenures together at Madurai Kamaraj University, my father in biochemistry and my mother in molecular biology. Madurai was a quaint temple town in the state of Tamil Nadu, located at the southern tip of India, far away from the bustling metros. I grew up in this simple town, protected by loving grandparents and parents, and never knew the outside world until I graduated from school. This was in the early 1970s, when Biotechnology was still an unknown field in India and not yet recognized as a discipline of its own in the biological sciences. My mother, Dr. Kunthala Jayaraman, was one of the pioneers in bringing this transformation to India, converting lab-based molecular biology and genetics into industrial applications. One of her first successful ventures was the mosquito

Fig. 1 Dr. Kunthala Jayaraman in 1969 as a lecturer in Madurai Kamaraj University

control program. Using techniques of cloning and expression in *E. coli*, she used a DNA fragment from *Bacillus sphaericus* to code for biocidal activity against mosquito larvae and successfully launched a mosquito control agent, running trials in the state of Kerala, a state in the southwestern part of India. She is the author of the book "Biopesticide update: BT based products." It was a rarity in that era for academicians to work with industry to launch commercial products and I am proud to say that she brought about that shift in mind-set in this community.

Thanks to her efforts, along with other peers of hers, the Indian government's newly formed Department of Biotechnology announced that biotechnology would become a separate science division, and Dr. Kunthala Jayaraman was invited to start one of the very first 'Center for Biotechnology' at Anna University in Madras, the capital of the state of Tamil Nadu, India. While this was a wonderful opportunity for her career, it meant moving away from her family to a new city. My sister was 16 years old and in her final year of school, while I was ten and studying in a convent school. Ours was a joint family and my grandparents opposed her wish to move to Madras. My father was torn between the two sides. Not daunted by anyone's objections, my mother made up her mind to accept the job in Madras with or without their support. One fine day, she bundled me into the car with a few suitcases, and we left for Madras, period! Keep in mind that in the era of the 70-80's, this was unheard-of in India. Grand daughter of a freedom fighter, daughter of a renowned lawyer and a violinist, I believe she inherited genes of bravery and intelligence. A few years back, I was doing a Google search on her name and, amazingly, I found a letter she had written in 1973 to Dr. James Watson, the Nobel Laureate, who co-discovered the triple helix structure of DNA. Fearlessly, she had written to him directly, seeking advice for her first course structure in Master of Sciences in molecular biology. I do not know if he ever replied! (Fig. 2).

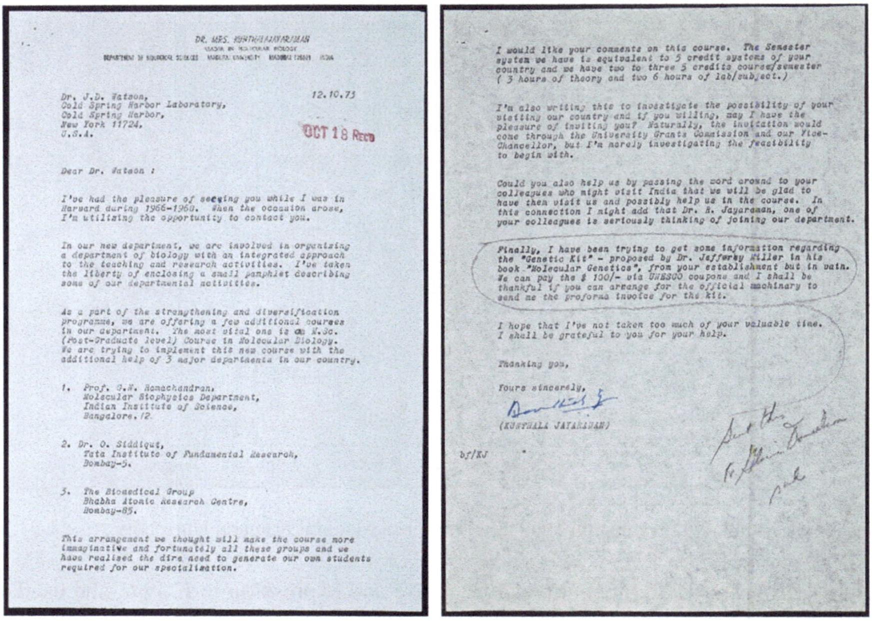

DR. MRS. KUNTHALA JAYARAMAN

12.10.73

Dr. J.D. Watson,
Cold Spring Harbor Laboratory,
Cold Spring Harbor,
New York 11724,
U.S.A.

OCT 18 Recd

Dear Dr. Watson :

I've had the pleasure of seeing you while I was in Harvard during 1966-1968. When the occasion arose, I'm utilising the opportunity to contact you.

In our new department, we are involved in organizing a department of biology with an integrated approach to the teaching and research activities. I've taken the liberty of enclosing a small pamphlet describing some of our departmental activities.

As a part of the strengthening and diversification programme, we are offering a few additional courses in our department. The most vital one is an M.Sc. (Post-Graduate level) Course in Molecular Biology. We are trying to implement this new course with the additional help of 3 major departments in our country.

1. Prof. G.N. Ramachandran,
   Molecular Biophysics Department,
   Indian Institute of Science,
   Bangalore. 12
2. Dr. O. Siddiqui,
   Tata Institute of Fundamental Research,
   Bombay-5.
3. The Biomedical Group
   Bhabha Atomic Research Centre,
   Bombay-85.

This arrangement we thought will make the course more immaginative and fortunately all these groups and we have realised the dire need to generate our own students required for our specialisation.

I would like your comments on this course. The Semester system we have is equivalent to 5 credit systems of your country and we have two to three 5 credits course/semester (3 hours of theory and two 6 hours of lab/subject.)

I'm also writing this to investigate the possibility of your visiting our country and if you willing, may I have the pleasure of inviting you? Naturally, the invitation would come through the University Grants Commission and our Vice-Chancellor, but I'm merely investigating the feasibility to begin with.

Could you also help us by passing the word around to your colleagues who might visit India that we will be glad to have them visit us and possibly help us in the course. In this connection I might add that Dr. R. Jayaraman, one of your colleagues is seriously thinking of joining our department.

Finally, I have been trying to get some information regarding the "Genetic Kit" - proposed by Dr. Jefferey Miller in his book "Molecular Genetics", from your establishment but in vain. We can pay the $ 100/- via UNESCO coupons and I shall be thankful if you can arrange for the official machinary to send me the proforma invoice for the kit.

I hope that I've not taken too much of your valuable time. I shall be grateful to you for your help.

Thanking you,

Yours sincerely,

(KUNTHALA JAYARAMAN)

bf/KJ

Fig. 2 Letter from Dr. Kunthala Jayaraman to Dr. James Watson

While working at Anna University in Madras (Madras was renamed as Chennai), she rose to great heights by running a well-funded department, conducting foreign collaborations, and developing high-quality PhD students, who worked for some of the largest biotech/biopharma firms across the world, such as Schering Plough, DuPont etc.. More importantly, she continued to work to improve the industrial angle of biotechnology by collaborating with varied companies to produce commercial products. Her department produced proteolytic enzymes used in the leather industry, enzyme manufacturing of Papain, and continued her earlier work with biopesticides, using BT-based carriers for environmentally friendly pesticides. During this time, my father, unfortunately, decided to stay back in Madurai with his parents, and it was a challenging time for my mother. She was juggling her work and international travel with trying to manage one riotous teenage daughter (me!) and another budding young woman in a tough medical school program. Her hard work was duly recognized, and she was elected as a Fellow of the National Academy of Engineering, India, shortly after she was elected a Fellow of the National Academy in Sweden (Fig. 3). This picture tells the story: a beautiful Indian lady in a sari (single parent of two girls) standing proudly and diffidently amongst an all-male international community.

From the early 1980s, my mother mentored dozens of women interested in pursuing a science career. She would grant admission to a few very bright young

Fig. 3 Dr. Kunthala Jayaraman elected as a Corresponding member of the Swiss Academy of Engineering Sciences

women into her PhD program based on their merit, and after 2–3 years, they would ask to drop out because their parents made an arranged marriage agreement for them. This would upset her terribly, not just for the loss of a PhD seat but also for the loss of a bright-minded young woman, who would now become a homemaker. She would try her best to counsel these women, sometimes meeting their parents and pleading with them to let their daughters finish their PhD. A Nature India article once quoted her, "One reason why very few women scientists are at the top is that several women drop out of the pipeline before or after completing PhD," says Kunthala Jayaraman, former Head of Biotechnology at Anna University in Chennai. "She attributes this loss partially to women being forced by their families to get married" [1].

As the progeny of two brilliant scientist-parents and with early exposure to my mother's fight for women in science and biotechnology, the will to succeed in this field is in my blood. I was more fortunate than most to have a patient father as my backbone and this dynamic lady as my role model. She would always tell me, "If you act like a delicate woman, you will be treated as one with no ability." From the time I learnt my numbers and could count, my Saturdays were spent in their labs. My mother would put me to work making agar plates, counting colonies, washing labware, etc. All around me, her students and PhD candidates would discuss their work and their ideas. This was my early music. The community spirit in her lab was so intense and close-knit that I am still in touch with many of her early PhD and Post-Doc students. KJ, as she was fondly addressed by everyone who knew her, passed away in 1998 while serving as the Dean of Academics at Vellore Institute of Technology. Though she is no more, she is still recognized as the Mother of Industrial Biotechnology Education and is remembered very fondly by her colleagues and peers [2] (Fig. 4).

Excerpt from an article authored by her student {2}:

"From Anna University to America and to Agriculture" Pandjassarame Kangueane* Bioinformation Volume 17 (1) Jan 31, 2021

Kunthala Jayaraman (KJ) is the mother of Industrial Biotechnology Education in the world who created the syllabus for it with a perfect mix of Science, Engineering and Technology at Anna University. She was extremely active, forceful, women of wonders, and an awesome dealmaker during post-Independence India. She appointed several faculty members who returned from USA after their doctoral/postdoctoral training. The subject was truly inter-disciplinary. She appointed people from various backgrounds in science and engineering, especially from Organic chemistry to Chemical Engineering so as to create a beautiful blend for Industrial Biotechnology (Figure 2 - see PDF version). She created "corpus funds" to support faculty members. The curriculum had fluid mechanics on one side and molecular biology on the other side. Thus, Industrial Biotechnology Education was novel to the world in 1992. Her birthday falls on the first of January every year and the New Year celebrations catalyzed creativity at her birthday parties.

Fig. 4 Excerpt from an article authored by one of her students

2 Biopharma: The Scenario in India

2.1 The Advent and Rise of the Indian Biopharma Industry

The Indian biopharma industry started as early as the 1970s with companies like Biocon Ltd. and Serum Institute of India Pvt. Ltd. venturing into biopharma products like vaccines and microbial products. While the global biotech revolution started in the mid-1980s, it was almost the 2000s before recombinant biotech products for medicinal use were first commissioned in India. The majority of these first-generation biotech products were still not indigenously developed in India but were collaborative projects with global companies, to be produced in India. A few examples of these collaborations include Biocon's tie-up with CIMAB, Cuba, in 2002; Dr. Reddy's Biologics joint venture with R-Pharm of Russia in 2010; and Lupin Biotech's collaboration with YL Biologics of Japan in 2014.

The Indian biotechnology industry formally began in 1986 with the establishment of the Department of Biotechnology (DBT) by the Ministry of Science and Technology. This was a first-of-its-kind initiative, even globally, dedicating an entire government department and subministry solely to biotechnology development. While India is a major player in the small-molecule industry, it was noticed that they were lagging behind in biotech. The government department was established to focus on biotech research and development with funds and support from the government. From being a basic research funding agency, by 2012 DBT had evolved into more of a partner for industries. DBT formed an independent, not-for-profit public enterprise termed Biotechnology Industry Research Assistance Council (BIRAC) to support this emerging biotech market. BIRAC has proven itself to be a powerful force for biotech and biopharma—financing start-ups and creating bioincubation centers for shared infrastructure. The Bioeconomy report, published annually by BIRAC, gives details of its achievements [3].

In 2021, BIRAC jumped into supporting the research, development, and production of COVID-19 vaccines with emergency funds. They financed an estimated $400 million (USD) for each of the vaccine companies like Serum Institute of India, Bharat Biotech, Biological E, Zydus, etc. This was an essential component of the success India had with the COVID-19 vaccine deployment. It is fitting to state here that BIRAC especially thrived under the woman leadership of Dr. Renu Swarup, Secretary, Department of Biotechnology (DBT), who was leading BIRAC from 2015 to 2021. Multiple accomplishments were done during her tenure (Fig. 5), and she dynamically led BIRAC with a clear agenda. A member of the task force on Women in Science under the Prime Minister of India, she also implemented the DBT Scheme on Biotechnology Career Advancement for Women Scientists—BioCare [4].

Fig. 5 BIRAC's achievement in biotechnology—India Bioeconomy report 2022

India's biotechnology industry is currently valued at US$80 Bn+ and is expected to reach US$150 Bn by 2025 [5]. With COVID-19 economy contributing to the spike in the growth in biopharma revenue in 2021, India's biotech sector grew from $70 billion (USD) to $80 billion (USD) in 1 year.

Biopharmaceuticals revenue is about 67% of the biotech industry (Biopharma at 49% plus the Covid Economy at 18%) (Fig. 6). Within this, therapeutics and vaccines contribute about 42% of the revenue. This sector is expected to double in 3 years, increasing its contribution to 75% of the total biotech revenue for India. Bioagriculture, Bio-Information Technology (Bio-IT), Contract Research and

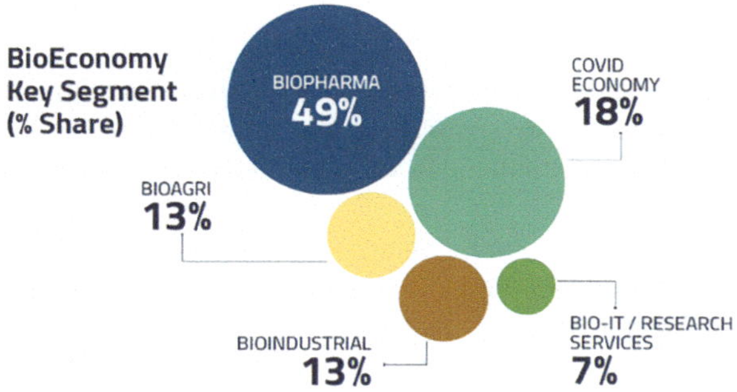

Fig. 6 Biopharma's share of India's Bioeconomy—India Bioeconomy report 2022

Contract Manufacturing, Biofuels, and enzymes (bio-industrial) are the other contributors to the biotech industry revenue.

Research & Development (R&D) spending in India for biopharma increased from $300 million (USD) to $1 billion (USD) in 2022. India spends around 0.8% of its gross domestic product (GDP) on R&D. While this is quite low compared to the world average of 1.8%, subsequent government investment in COVID vaccine development and production gave a much-needed boost to biopharma R&D investment. The Indian biopharma market needs this kind of large-scale funding influx into their R&D for all other Biologics, Biosimilars, stem cell therapy, adeno-associated viruses, and other new technologies in biopharma to rise faster in this competitive environment. It could be said that if the investment received during the pandemic for R&D, manufacturing, and QC equipment infrastructure of COVID-19 vaccine production is used wisely, it could be still be deployed for multiproduct facilities and, hence, newer vaccines or Biologics.

2.1.1 Top 5 Unique Features of the Biopharma Industry in India

1. Family-Run Businesses

Looking at the Indian biopharma industry from an helicopter field of view, it is a unique situation in India that a greater majority of the biopharma companies are family-run businesses, while global biopharma companies are usually publicly traded, listed corporations. This gives the Indian companies the freedom of quick decision-making and on-the-ground leadership. In vaccine companies like Serum Institute, Biological E, Bharat Biotech, or Biosimilar companies like Intas Pharma, Zydus [6], Lupin, or Dr. Reddy's, the top management is typically a father-son (in one case, father-daughter) legacy. Capable of quick yes or no decisions, when and where it matters, prompt reactions and pivots can be taken to adapt to the market situation. Normally, such decisions will go through long-winded processes with management board approvals and stock market considerations. It is my opinion that this is one of the main factors India was successful in COVID-19 vaccine production at such a rapid pace and at a scale sufficient to supply to the WHO countries.

2. Chemistry and Mathematics

India has always excelled in chemistry and mathematics. The chronology of Indian mathematicians and physicists spans from the Indus Valley Civilization and the Vedas, from Aryabhata to Ramanujan to Shakuntala Devi (known as the human computer).

Aryabhatta was the first major mathematician-astronomer during 470–550 BC. His book *Aryabhatiya* introduces arithmetic, algebra, plane trigonometry, and spherical trigonometry to the world. It also contains continued fractions, quadratic equations, sums-of-power series, and a table of sines. He also clearly defines the "place-value" system in mathematics. Srinivasa Ramanujan was a nineteenth-century mathematician who contributed vastly to the theory of numbers. Even more recently, we have Dr. Shakuntala Devi, the mental whiz, who could give the 23rd root of a 201-digit number in 50 seconds! In physics, Dr. Chandrasekhar shared the 1983 Nobel Prize for physics with William A. Fowler for "theoretical studies of the physical processes of importance to the structure and evolution of the stars" [7]. The Chandrasekar Limit changed the outlook in astronomical studies. Similarly in chemistry, which was referred to as "Rasayana" [8] and practiced since the sixteenth century AD, it was an ancient art amongst Hindu gurus, whether it was Ayurveda or related to human health, or metal and nonmetal chemistry. Chemistry is one of our heritage traits.

India is referred to as the pharmacy of the world and has a large pharma generics business. India is the world's largest provider of generic medicines by volume, with a 20% share of the total global generic market. It is also the largest vaccine supplier in the world by volume, accounting for 50–60% of the world's vaccines by volume [9]. By revenue, India's pharma industry is estimated at $42 billion (USD) and ranks 3rd worldwide for pharmaceutical production by volume and 14th by value [10]. There is a focus on chemistry in basic education because of the history of India and also because of the job opportunities in pharma companies. Naturally, this has benefited the biopharma industry, where strong biochemistry backgrounds are essential. Like reverse engineering, "reverse characterization" is the art and science of making sure the Biosimilar characteristics exactly match the innovator biologics—not worse but also not better! This is where the Indian scientists shine. Even a slight variation could make the molecule not a Biosimilar anymore. In numerous cases, the Biosimilars may have actually shown better characteristics than the innovator molecule, obviously due to the advancements in manufacturing technology, but no, that cannot be! Precision Protein Pairing - this is the skill of our Indian biochemists.

3. Advantage Pharma

Very rarely is there a "pure-bred" biopharma company in India, except maybe for some of the vaccine manufacturers. The investment or idea to diversify into Biopharmaceutical production, leading to the start of a Biologics facility, is usually from a large pharmaceutical company, who is already a well-established player in the world arena of small-molecule generics. India has a strong domestic pharmaceutical industry comprised of ~3000 drug companies and ~10,000 manufacturing units.

Some of the top generics players are Sun Pharma, Lupin, Dr. Reddy's, Alkem, Cipla, and Intas. When they decide to diversify into biopharma, investment is high and readily available, and, most importantly, there is already a well-established global marketing channel, selling generics to the world, which can be repurposed to sell Biosimilars. This was a huge advantage over other countries because global marketing costs can make or break such initiatives.

However, as the complexity of R&D and manufacturing of Biologics was ill-understood by the investing pharma giant, the expectations for an early return on investments like in the small molecule market clearly couldn't be met and quite a few Biipharma ventures quickly folded up operations, resulting in projects that were dropped prematurely. The ones who sustained the early challenges and persevered are now not only thriving but are global leaders in of Biosimilars sales. Some examples of successful Indian Biopharmaceutical companies are Biocon, Intas, Zydus, Dr. Reddy's, Lupin, Enzene, Hetero, etc.

Recent years have seen the rise of start-ups in the biopharma space with investments in manufacturing facilities. Companies like Procell, GenBio, Theertha, Provis and others have investors from private finance and venture capitalists (in contrast to the big pharma investment from well-established Indian pharma companies) and have had the sense to draw the ROI timelines more accurately as compared to early biopharma investors. These companies tend to have a "Go CMO" (Contract Manufacturing Organization) strategy. From the beginning, they are establishing full-scale manufacturing units equipped with single-use, multiproduct facilities and contracting work out to global Biologics companies, while parallelly working on their own R&D pipeline. This is a win-win situation, as the global Biologics companies can reduce their manufacturing costs, and the start-ups have a self-sustaining facility ready to use when they get their own Biosimilar out of R&D.

4. Brain Regain

As opposed to an earlier era of brain drain, brain regain is happening in India now. For the previous two decades, India's scientists left the country in droves to pursue their scientific careers abroad. Well-trained PhDs in biological sciences would get a Post-Doc abroad in the United States or Europe and never return. Quite a few of them are successful biopharmacists in global companies. Maybe due to lucrative compensation packages, state-of-the-art facilities, and, in general, a smoother lifestyle than in India in the twentieth century. Things have changed dramatically; in the last decade, the biotech brain drain is now becoming a biotech brain regain with several Biotechnology scientists , who studied and worked in the US, Eutope, Korea and Japan are returning back to the country with enriched knowledge and experience. These scientists have been lured back by good government grants, such as packages for setting up new labs, young scientist grants, and more such government funds. Industrial experts have much better compensation packages now with India's biopharma companies and are wooed back with fancy incentives. Of course, the critical pull to come back is India's fastest-growing GDP growth rate (6.9% in the first quarter of 2023), open economy, much-developed infrastructure, and the added bonus of family connections and familiar culture.

5. Scale at Cost

The concept of scale is something India understands very well. With a population of one billion people, scale is a necessity, and with 34% of the population below the global poverty line, cost is another essential. Take the Jio cellular service for example, where cellular phones were given away at $20 (USD) per year—just $20 and data at even less! The Jio model worked so well because of the sheer volume of people subscribing. India is now the largest consumer for apps like WhatsApp, Facebook, etc., each with a billion users! With an active user base of 0.5 billion WhatsApp active users, India easily surpasses Brazil at 0.1 billion and the United States at 91 million active users. Similarly, there are more than 380 million active Facebook users, compared to the United States at 180 million and Brazil at 120 million [11].

Closer home, the vaccine industry is an even better example, as India not only manufactures enough for the Indian population (25 million babies in a year) but also supplies 50–60% of all vaccines to the world by volume [10]! The supply of traditional vaccines to the World Health Organization (WHO) countries is a perfect example of scale at cost. You had to produce enough vials of vaccines to vaccinate 60% of the world's children every year for MMRV (measles, mumps, rubella, and varicella (chickenpox), pneumonia, etc., and for UNICEF, WHO, and national governments, who would only pay a highly subsidized amount for each dose. In spite of the low sale price to these entities, this has still been a profitable business for companies like Serum Institute of India and Biological E, who have learnt to perfect the art of scale at cost. One Indian rupee ($0.02 USD) a dose still means 1 billion rupees when you produce at that scale!

India produces nearly four million doses of COVID-19 vaccines a day, for a total of 1.45 billion doses in 2021 [12]

The WHO reports that in India, from 3 January 2020 to 25 January 2023, a total of 2,203,275,159 COVID-19 vaccine doses have been administered. All 2.2 Bn doses—nearly four million COVID-19 vaccines a day in 2021—were made in India and exported to African and other neighboring countries. Serum Institute alone was producing 100 million COVID-19 vaccine doses a month. This is massive scale at cost!

This concept, if successfully applied to biologics like mAbs (Monoclonal Antibodies), will open up a huge, needy market and remove the current situation where Biopharmaceutical therapies are available only for the rich and insured. We are not there yet, but that should be the goal.

2.2 *Biosimilars: Their Journey in India*

Established guidelines for biosimilars or "similar biologics" came into existence in India only in 2012. My career in biotech started in 2012, and I can trace the evolution of biotech in synchrony with the growth of my career in this industry. Granting regulatory approvals for "similar biologics" requires much more data than a simple generic chemical drug application. According to GABI (Generics A and Biosimilars Initiative), the first biosimilar ever approved from India was for the Hepatitis B vaccine in 2000. We have come a long way since! Filigrastim from Intas/Accord Pharma was the first biosimilar ever accepted in Europe in 2014. This was the much-needed boost awaited by the Biopharmaceutical industry and opened the floodgates favoring the entry of Indian biosimilars into global markets.

2.2.1 What Are Biosimilars?

While there are multiple descriptions and explanations, the one I prefer is from Pfizer: "A biosimilar is a biologic product that is approved based on demonstrating properties that are highly similar to an already Food and Drug Administration, USA (FDA) approved biologic product known as the innovator or reference biologics. A biosimilar has no clinically meaningful differences in the safety or effectiveness from the reference product" [13].

The majority of biosimilars are monoclonal antibodies (mAb) products, including therapeutic monoclonals, fusion products, antibody fragments, antibody-drug conjugates, etc. mAbs are immunoglobulins composed of two light chains and two heavy chains connected by disulfide bonds. The mAb structure can be divided into three fragments: two antigen-binding fragments (Fab), each with one light chain, and one crystallizable fragment (Fc) containing the heavy chains. The Fab chain binds to antigens (substances foreign to the immune system, such as bacteria or viruses) (Fig. 7). More details can be found elsewhere [14].

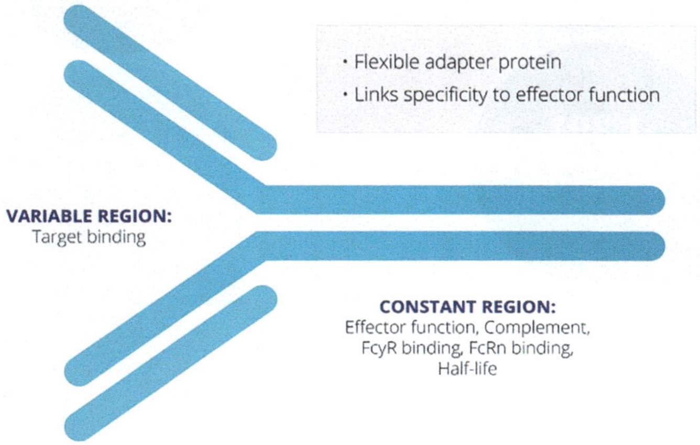

Fig. 7 Basic antibody structure-function [15]

Innovator drugs (or reference products) go through quite a long and tedious process of clinical trials to get approval to market. However, the FDA approves biosimilars through an abbreviated pathway (Figs. 8 and 9) [15].

The goal of a biosimilar development program is to demonstrate biosimilarity between the proposed biosimilar and its reference product, not to independently establish the safety and effectiveness of the proposed biosimilar. This means that

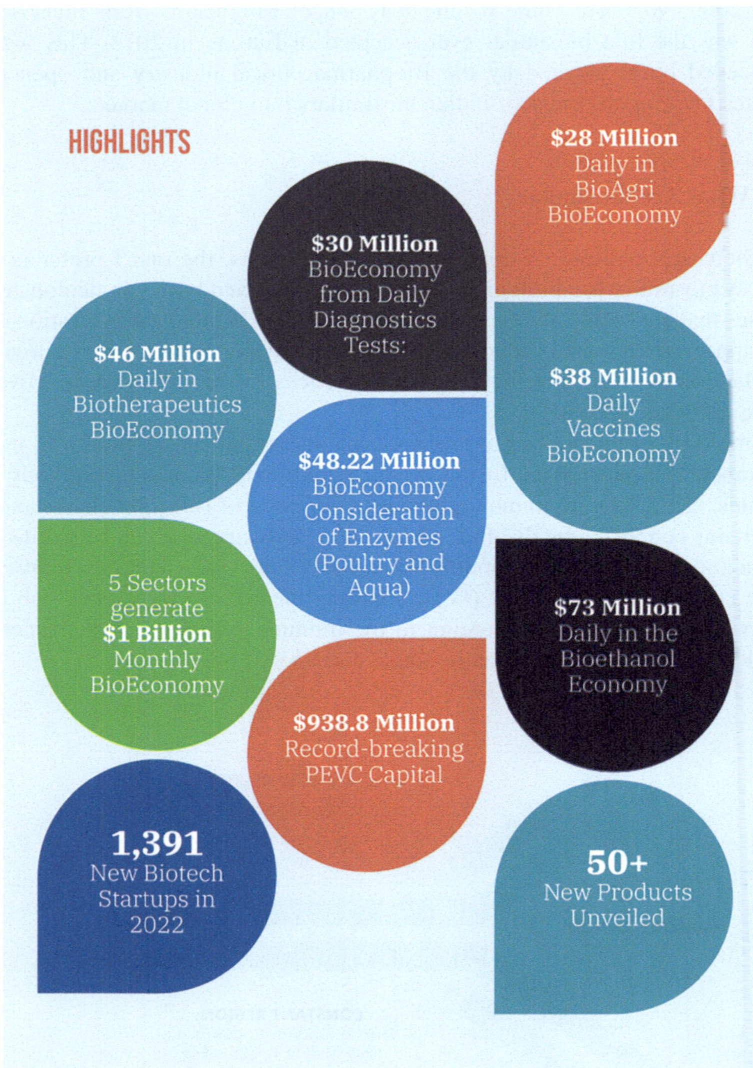

Fig. 8 A biosimilar approval process from the FDA [14]

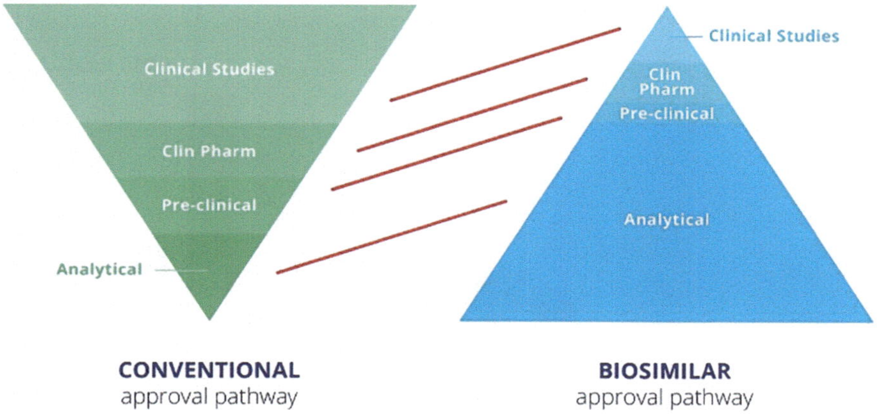

Fig. 9 Regulatory requirements for approval of a conventional biologic versus a biosimilar [15]

biosimilar manufacturers do not need to conduct as many expensive and lengthy clinical trials. The focus is more on analytical studies demonstrating that the product is highly similar to the reference product, animal studies, and a clinical study [16]. The original biologic is protected until its patent expiry, and this is when the biosimilars jump in. Let me narrate the Humira® story for a better understanding of the disruption biosimilars can cause.

2.2.2 The Curious Case of Humira Abbvie (Fig. 10)

Humira is an original biologic drug used primarily for the treatment of rheumatoid arthritis. Humira, produced by AbbVie (after years of R&D money, expertise, and of course no competition), is priced at $7389 (USD) for two subcutaneous kits. Humira, whose biological name is Adalimumab, is a monoclonal antibody. The patent filed by AbbVie expired in 2018 (principal US patent).

AbbVie rushed to get this extended till 2023 and made multiple settlements with biosimilar manufacturers to postpone the launch of a biosimilar version of Humira from these companies. Care to guess why? Global revenue of Humira reached $20.7 billion (USD) in 2021 [18]. Any biosimilar of Humira is bound to be launched at a quarter of the price of the innovator drug. A biosimilar could do two things: (1) Eat into the revenue from Humira sales and (2) force Humira to drop prices to keep up market share. And that's exactly what has started to happen. In April 2023, AbbVie's stock crumbled 8%! The revenue downfall of more than 25% was in reaction to Amgen's biosimilar named Amjevita, which was launched halfway through the first quarter. And this is just the first biosimilar to hit the US market; there will be more [19].

Several Indian biopharma companies started their research and development on biosimilar versions of Humira as early as 2012, with the hope to be the early launcher of a Humira biosimilar. There are reported to be seven different biosimilars

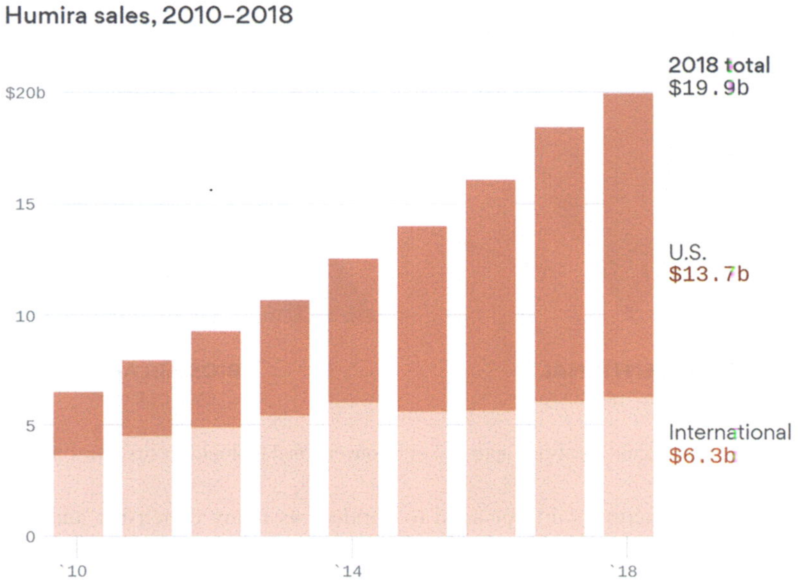

Fig. 10 Humira sales approach $20 billion in 2018. Data—Company filings [17]

currently produced by India, jostling for share. In 2014, Zydus Cadila was the first to launch Exemptia, and they could only market in India and Europe as AbbVie still had a patent in the United States. In 2016, Torrent launched Adfrar, followed by Hetero in 2018. Close to launch were Glenmark, Cipla, and Emcure. Interestingly, Indian players appear to have aligned the prices of their adalimumab biosimilars in the region of around INR22,000–25,000 (US$320–364), except for the Cipla brand, which is priced at around INR16,072 (US$270), details on online pharmacy sites in India indicated. Compare this with the $7389 USD for two kits by AbbVie.

As of today, there are already eight globally produced biosimilars of Humira approved by the US-FDA in 2023, including Viatris, Samsung, Pfizer, Sandoz, Boehringer Ingelheim, Amgen, Coherans, and Fresnius Kabi. As mentioned above, just one of these launches from Amgen has already created havoc on Humira. This is bound to get more competitive and economical as more competitors enter the fray.

How are the Biosimilars companies able to do this? How are they able to market and sell a unique, life-changing biologic drug at one-tenth the price of the innovator?

- Lesser spend on R&D: The protein or antibody of interest is already known and characterized and now only needs to be resynthesized or recreated from a clone. Hence, there is no lead screening time or cost involved to identify the right drug.
- Faster time to market: Pre-established therapeutic effectiveness of the innovator or reference product means there is no need to establish or evangelize the product again. Less stringent clinical trials requirements.

- Modern technologies: Cost of production decreasing rapidly over the years with technology like multiproduct/single-use facilities, purification costs, etc., contributing to better purity, efficiency, and throughput.
- Readily available marketing avenues are well-established for generics. This is specifically for Indian biosimilar players as they have an established network globally for selling generics (small-molecule drugs), which can be leveraged effectively for biosimilars without additional investment.

3 The Evolution of the Biopharmaceutical Industry in Parallel with My Professional Evolution

2010 is when Indian biopharma companies came of age. There was excitement in the air as press releases abounded about various companies starting their biopharmaceutical programs. At the time, I was the Senior Marketing Executive at GE Healthcare Life Sciences, based in Bangalore, India, and could sense the change in the atmosphere. GE Healthcare Life Sciences provided instrumentation and consumables needed by the biopharma producers, from R&D scale to pilot and production scale capacity. We needed a plan, a think-big strategy to lead the next decade, staying ahead of the competition and, all the while, being available to help our customers win in the biopharma world. Zydus Healthcare became one of the first biopharma companies in 2014 to invest in large-scale manufacturing for monoclonal antibodies. They invested in 5000 L stainless steel bioreactors, the largest in India for mAbs to date, and they needed downstream purification to match the scale. The deal structure was such that, while they needed hardware (instrumentation) for about 1 million USD, they would need purification media (consumables) worth 3–4 million USD over the next 3 years. We were able to structure the deal in such a way that we could show them savings over 3 years as a total cost of ownership. It was close, though, and had we focused on just the initial instrumentation pricing, we would have been twice the price of competitors and lost miserably. This was a valuable lesson for me on the value of hardware (instruments) versus consumables (purification media) and the nuances of profitability and pricing mix. I realized how critical it was to have consumables in the equation always. This was the initiation of my entry into the commercial world of biopharma.

We launched a country-wide purification media start-up kit program targeting mAbs (monoclonal antibodies) purification. The media kit contained small quantity samples of assorted purification media for the customer to try out on their own molecule, totally for free. We realized that lab scientists, who work on these protocols, were quite junior and less experienced, so we added another layer to the free resins kit. We included the time of our trained application scientists to help the lab scientist with their trials. This was the clincher! A freebie goodie, in general, is usually not respected by any of us and usually lies unnoticed in a corner of our home (or lab). With a person attached to it—a live person sitting in your lab and helping you use the freebie—that had a different impact! Our application scientist would carry this media kit with them and spend 2–3 weeks in the customer's lab, experimenting and optimizing in the presence of the customer. This initial investment worked beautifully, and within a span of 3–4 years, our purification media was "designed-in" to almost 90% of the mAb purification processes in R&D in India. This was a substantial achievement, and we received global accolades for it, not to mention profitable deals.

When a purification media is optimized and "built" into your R&D process, you are already establishing certain biosimilar analytical characteristics of the end product and comparing it with the innovator. When a high similarity is identified, the specific molecule and process are selected for further studies and scale-up. If successful, a dossier is submitted to the FDA and other regulatory agencies for approval to proceed to clinical trials. After this point, no component in the entire process can be changed, considering the odds of how it may affect the end product and its characteristics. The identified product and process now go into a pilot-scale mode (higher capacity and higher output) and from there to a process scale mode. In this entire scale-up process, they must now buy GE's purification media. Every R&D insert of our media, and our application person, was celebrated and recognized. As we learned from the Zydus incident of 2014, if our media was a mandatory purchase commitment for them, we could bundle and package the instrument, service, and media into a very winning deal. I am proud to say that today, more than 80% of India's mAbs programs and more than half of our vaccine products use GE's downstream hardware and media.

The next focus was on the characterization of mAbs. India has the tendency to wait and watch: Is a new characterization technique for mAbs approved by the FDA or EMA? Is it yet part of any other company's release test or FDA dossier? Is it used by innovators for their QC release protocol? Is it a "must-have" or a "good-to-have" technique? We had an excellent product in GE Life Sciences that we knew was the future, selling like hot cakes in western countries, while Indian biopharma was not yet ready to adopt it—they were waiting and watching.

How do I speed this up? The marketing team I was now heading developed the novel concept of "Evolutions"—a series of customer seminars targeting prospective Biacore™ buyers, where we got the world to come to India.

Biacore, based on Surface Plasmon Resonance (SPR) technology, was one of the key techniques identified for analytical characterization and comparison of a biosimilar to a reference product or innovator. As mentioned earlier, the Biosimilar approval pathway was heavily based on analytical characterization, and Biacore/SPR was one of the proven techniques used by global pharma giants. Biacore's data was trusted by the US-FDA as part of the drug approval process. While it was not yet officially made a part of the FDA guidelines for Biosimilars in 2010, we knew it was only a matter of time. India was taking its time, adopting a wait-and-watch attitude, and we needed to speed this up!

We got industry experts from global biopharmaceutical companies to come here and speak about the acceptance and importance of this technology. We identified Indian-origin scientists in the United States and Europe working with multinational pharma companies on this technology, who were quite happy for an opportunity to visit India and combine their professional talks with their personal family reunions. This was a year of hectic travel, accompanying these illustrious foreign scientists across the length and breadth of India, but it was also glorious fun interacting with international researchers and pharma managers. We broke the inertia in the market, and within a few years, we sold Biacores to these Biopharma companies much before it became an FDA regulation and much before the competition even woke up to it (Fig. 11).

Our next opportunity came when competition was increasing in the market, and we needed a different strategy to stay ahead. Researching the Biopharma trends in the market, we understood that the FDA was promoting the QbD concept (Quality by Design) for biopharmaceuticals, and this was a new area for us. We decided to launch the QbD series of workshops for our biopharma customers—a 2-day workshop series with lectures given by the top best and brightest minds in the industry and academic—with an aim to elevate our customers to the next level of thinking. The workshop covered trendy and serious topics around CQAs (Critical Quality Attributes), DoE (Design of Experiments), IPQC (In Process Quality Control), PAT (Process Analytical Technology), Control strategy, and Continual improvement.

Fig. 11 My marketing team and I receiving an award from Terri Bresenham, CEO of GE Healthcare India, for Evolutions, 2012

This was the Six Sigma equivalent for biologics, and our customers loved it. We once again elevated ourselves to a more elite rung of the ladder.

By now, I had traveled the length and breadth of the country running these programs and meeting my biopharma customers. While professionally stimulating, it was a tough time in my personal life, managing two young school-going boys, managing unstable nannies, and an overburdened husband. Travel within India was not easy either for a woman, with a scarcity of ladies' restrooms during road trips, safety issues in cabs, and sometimes a bad tummy from a street food diet. My husband was a huge support for me, and in retrospect, his career got stagnant as he had to restrict his own travels and projects for me and the kids. His physical presence was essential at home when I traveled. The strength behind me was, however, the spiritual presence of my mother. I could always feel her urging me on, pushing me to greater efforts, and encouraging me when I wanted to give up. There were instances that were traumatic, and sometimes I would even want to quit my job. For example, in 2005, I was in Hong Kong for an internal audit training course when my 5-year-old son had an epileptic attack! I felt so helpless. There was nothing I could do for him from there. When I got a call from the Hong Kong office front desk, I couldn't breathe. Of course, my family managed the situation quite well, which could have been a temporary toddler issue, but until I got home 24 hours later, it was terrifying. On the lighter side, when Michael Jackson died in 2009, my eldest son, a tough teenager, musician, and huge MJ fan, was inconsolable. As always, I was traveling and all I could do was cry along with him over a terribly bad landline connection. My sons are close to me even today, and I can feel no resentment from them, only pride in their mother's work. I could deal with all this and more just because I loved this biopharma industry and believed in the job I was doing. In the end, that is what matters.

In 2014, I transitioned from a marketing role to a commercial role as the General Manager of Western India for GE Healthcare Life Sciences. Things got harder—or that was what I thought in the beginning. I moved my entire family from Bangalore to Bombay with my husband and youngest son tagging along. My son was 14, a typical teenager and was being upended from his familiar life. He had to join a new school, learn a new language and try and make new friends, from a very different culture. He hated me - or so I thought.

In my new role, I was dealing with a large team of sales and application people, a calcitrant son, a new home, and city. It was a nightmare for a while. My boss and mentor at GE Healthcare, Mr. Anurag Gupta, called me once and told my very simply, "Everything happens for a reason." Six months later, these reasons were clear. The western biopharma industry was booming. West India is where the largest biopharma companies are located like Serum Institute, Zydus, Intas, Lupin, and Sun Pharma, to name a few. I was, once again, in the right place at the right time. We worked on massive production-scale biopharma projects with our customers. At Cipla Biotech, we designed, sold, and installed their 3 × 2000 L bioreactors and all the downstream purification equipment for this scale. At Serum Institute of India, we collaborated with the customer team on their conjugate protein (a carrier protein for vaccines) project for the installation of multiple instruments in their downstream hardware requirements, and their process optimization with our purification media.

These were all projects with millions of dollars in revenue and changed the scale of business. We could discuss FlexFactory (GE's product offering of a centrally automated, integrated, flexible biomanufacturing platform) with our customers, looking at their project plans for entire facilities and coming up with custom solutions. Around this time, GE ventured into single-use bioreactors, upstream cell culture media, continuous chromatography, inline dilution, and multiple other modern technologies. It was a true opportunity to be in the midst of new-age technology, learning every day. I was exposed to the highest leadership in GE, and I traveled across the world, taking our customers to demo sites. I could not have had a better time in life.

In addition to the driving sales growth, I also learned to be a leader with the responsibility of motivating and driving my team to execute these large-scale projects. My team of 12 young men, consisting of sales specialists, application specialists, and service engineers, came to know me well, and we developed a great bond that still exists after a great many years. I learned to connect with them, find what makes them click, and instill in everyone a sense of pride and teamwork.

On the personal front, the situation improved quite a bit with our Mumbai lifestyle. My son loved his school life, and with their style of teaching, he blossomed from a shy introvert into a confident young man—class valedictorian, football captain, and school president! My husband loved Mumbai and started a variety of business initiatives there. Things were so much better—everything happens for a reason!

As Biopharma boomed, so did my career, and I am always grateful to GE for shaping me the way I am today (Figs. 12 and 13). An important part of the joy and pride I took in my work was my customers. I should say that I had a slight advantage as a woman. My customers' instinct was to make sure of my safety and comfort. I would be the only lady in the boardroom negotiating a deal or in their factory with a hardhat, supervising the installations. India is a matriarchal society; therefore,

Fig. 12 Visit to Amgen facility in Dublin

Fig. 13 With Kieran Murphy, then CEO of GE Healthcare Life Sciences—2016

men do not raise voices against women in general. This can be helpful when things go awry.

Even in huge negotiations, they remained polite and well-mannered because there was a tiny, 5-foot woman in their midst. The respect they showed me was a large contributor to why I loved this job. Obviously, they could also see that we were dedicated and that we genuinely wanted them to be successful, judging by the long hours that my team spent in their labs solving their issues, and by me sitting there in the field along with the team to encourage them and be available for quick decisions.

GE is the best place for women professionals, and I have seen no bias or hesitancy in promoting women. I was fortunate to be with GE for 17 years, surrounded by the best minds in Biopharma and encouraged by impressive women mentors like Terri Bresenham and Ipsita Gupta.

I moved on in 2017 to Beckman Coulter, then Cochlear, and now I continue my journey in the Biopharma industry, dealing with Genomics, Proteomics, Cell culture, and Imaging instrumentation at Spinco Biotech, Chennai, India. As you can see from the picture below (Fig. 14), things have not changed much, it is still a man's world in the Biopharma business!

Fig. 14 With my current team of dynamic young men, receiving our award for excellence is sales performance 2021 received from Molecular Devices, a Danaher company

4 Women in India's Biopharma Industry

4.1 Women Influencers in Indian Biopharma

It is quite providential that Indian Biopharma has quite a few women influencers—strong and idealistic women. Top CxOs of biopharma companies such as Biocon, Biological E, and Vimta are women. While this is just a handful, they still have managed to inspire younger women students to consider biotechnology as their career choice.

Dr. Kiran Majumdar Shaw needs no introduction as a pioneer in biopharma. Women like Dr. Mahima Datla, CEO of the second-largest vaccine manufacturer, Biological E; Haritha, CEO and MD of Vimta Labs, a top pharma and biotech CRO; Dr. Suma Ray, leading the plasma production unit at Intas; Dr. Asha Mallya, a stalwart Quality Control head at Serum Institute; and Dr. Anita Krishnan and Dr Laxmi Adikari of Biocon Biologics have established themselves in the biopharma world in India and show their commitment to this industry with successful ventures.

Almost all of these women are from the 60s and early 70s generation, and it is indeed troubling that the current generation of women is struggling to make it big in the biopharma industry. There are quite a few options for postgraduate education in biotechnology, biotech finishing schools like BVarsity and Biocon Academy, and internship opportunities for the younger generation, but a lack of rising talent at the middle or senior levels remains a cause for concern. Diversity hiring, grooming, and mentoring are still very new in India, and this is where the focus needs to be.

4.1.1 Young Women in Biotech

On the positive side, more young women are taking up the brave entrepreneurial route in India across all industries, and biotech is no exception. We are seeing bright, young women with good market-worthy ideas come up quite fast in the biotech industry in the past 3–4 years. The funding opportunities for them are vast, including BIRAC, biotech venture capitalists like Menterra, and private funding options. Bioincubator parks and entrepreneurship promotion facilities like CCAMP (ccamp.res.in) and VCR park (vcrpark.com) provide them the necessary state-of-the-art infrastructure for the initial R&D with minimal investment. Young women are, however, preferring low-investment areas in biotech like med-tech, healthcare AI, diagnostics, enzyme manufacturing, etc., and biopharma is still a huge leap for them because of the initial high investment risks.

I believe that the gap lies in the switch and jump of women scientists from biotechnology basic research to hardcore biopharma R&D and manufacturing. India's biotech academia has always been dominated by brilliant women, and such is the scene now too. However, enabling and empowering them to jump into translational research and then industrial-scale production has been a struggle. About 600,000 biotechnology graduates are generated every year in India, 40% of whom are women, but the way forward is unclear to the multitude of them as they do not have the right skills necessary to enter the biopharma industry. A change of syllabus in the graduate biotech schools is urgently needed, with the curriculum moving from standard cloning and microbial assays to protein manufacturing—biofermentation, protein purification, and protein characterization. The initial plan to start a biopharma R&D venture can be quite daunting, despite available resources.

More education and investments are needed in novel areas of biotech such as synthetic biology, AI-based drug modeling and design, and immunotherapies. CAR-T therapy (Chimeric Antigen Receptor T-cell therapy) is going to be the next big wave in biopharma, and this is, in my opinion, an open playing field for India. We should get in there as innovators with our own CAR-T therapies and solutions. This is one area where we need not be followers anymore. It is encouraging that two of India's top hospitals have started clinical trials this year. Quick and appropriate investment by the government and industry at this time should be the next big agenda.

References

1. Jayaraman, K. S. (2008, March 14). India woos its women scientists. *Nature India* (nature.com).
2. Kangueane, P. (2021). From Anna University to America and to agriculture. *Bioinformation, 17*(1), 29–36. https://doi.org/10.6026/97320630017029
3. Narayanan, S. *India bioeconomy report 2022*. Birac. https://birac.nic.in/webcontent/1658318307_India_Bioeconomy_Report_2022.pdf
4. BioVoice. (2021, September 2). Dr Renu Swarup becomes the first ever woman secretary of DST. *BioVoiceNews*.

5. Malik, R. *Biotechnology industry in India – Biotech sector analysis*. Invest India. https://www.investindia.gov.in/sector/biotechnology. Accessed June 6, 2023.
6. The Economic Times. (2017, July 12). *Pankaj Patel steps down as Cadila MD, son Sharvil to take over.* https://health.economictimes.indiatimes.com/news/pharma/pankaj-patel-steps-down-as-cadila-md-son-sharvil-to-take-over/59570701
7. The Nobel Prize. *Nobel Prize in physics press release 19 Oct 1983*. https://www.nobelprize.org/prizes/physics/1983/press-release/
8. Goyal, M. A. (2018). Rasayana in the perspective of the present scenario. *Ayu, 39*(2), 63–64. https://doi.org/10.4103/ayu.AYU_300_18
9. FICCI. (2018, February 9). *Trends & opportunities for Indian pharma 2018*. https://static.investindia.gov.in/s3fs-public/Trends%20%26%20Opportunities%20for%20Indian%20Pharma_0.pdf
10. IBEF Pharma Report. (2023, February). *IBEF Presentation*.
11. Ceci, L. (2023, February 17). *Number of WhatsApp users in selected countries worldwide as of June 2021*. Statista. https://www.statista.com/statistics/289778/countries-with-the-most-facebook-users/
12. Press Information Bureau. (2022, July 19). *India bioeconomy report 2022*. Press Information Bureau (pib.gov.in).
13. Pfizer. *Biosimilar basics, what is a biosimilar? Biosimilars frequently asked questions*. Pfizer Biosimilars. https://www.pfizerbiosimilars.com/faqs#:~:text=A%20biosimilar%20product%20is%20a,effectiveness%20from%20the%20reference%20product. Accessed June 8, 2023.
14. Goli, V. A. R., & Butreddy, A. (2022). Biosimilar monoclonal antibodies: Challenges and approaches towards formulation. *Chemico-Biological Interactions, 366*, 110116. https://doi.org/10.1016/j.cbi.2022.110116
15. FDA.gov. (2022, December 13). *Biosimilar basics for patients*. Review and Approval | FDA.
16. Isaacs, J., Goncalves, J., Strohal, R., et al. (2017). The biosimilar approval process: How different is it? *Considerations in Medicine*, 1. https://doi.org/10.1136/conmed-2017-100003
17. Herman, B. (2019, January 25). *Axios health Humira sales approach $20 billion. Humira sales approach $20 billion (axios.com)*.
18. Dunleavy, K. (2022, February 22). *Humira rings up $20.7 B in 2021, but AbbVie still mum on post-biosimilar expectations*. Fierce Pharma.
19. Gatlin, A. (2023, April 27). AbbVie's Humira downfall is swift as biosimilars hammer away at sales. AbbVie stock crumbles as Humira biosimilars take a 25% cut. *Investor's Business Daily* (investors.com).

Contribution of Mexican Women in the Biopharmaceutical Field

Alicia Soto

Contents

Acronyms

CAR-T	Chimeric Antigen Receptor T-cell
COFEPRIS	Federal Commission for Protection against Sanitary Risks
CONAHCYT	National Council of Humanities, Sciences and Technologies
ENOE	National Survey of Occupation and Employment
ICH	International Conference on Harmonization
INEGI	National Institute of Statistics and Geography
NOM	Official Mexican Standard
SSA	Secretary of Public Health
STEM	Science, Technology, Engineering and Mathematics
Tuxtlas	Geographical region located in Veracruz
UANL	Universidad Autónoma de Nuevo León
UNAM	National Autonomous University of Mexico

A. Soto (✉)
Agilent Technologies, Mexico City, Mexico
e-mail: Alicia.soto@agilent.com

S. Taylor, M. Campbell (eds.), *Women in Biopharma*, Women in Engineering and Science, https://doi.org/10.1007/978-3-031-80348-2_9

1 Promoting the STEM Careers for Young Women

In Mexico, the economic, political, and cultural backwardness over decades has caused the gender gap between men and women to be very evident at the family level; young women in middle and high school are not very inclined to pursue careers in the areas of science, technology, engineering, and mathematics because these are considered careers with difficult access, high costs, or because of comments from parents or relatives. These women think these are careers where there is no future that ensures a job, which allows them to have financial stability for future life projects.

In cities, there may be greater opportunities to promote STEM careers, but how can we reach those middle and high school girls in rural towns or small towns to choose these careers? We are faced with the fact that once they choose a career and achieve the academic degree, they must find a job that is related to these careers, one that allows them to have a balance between the development of personal and work life. But also, one that does not lead to the abandonment of the profession when they get married, become mothers, or take care of a family.

We could start listing some of the skills that are required in Science, Technology, Engineering, and Mathematics careers, such as the following:

- Problem-solving
- Ability to learn and analyze
- Critical thinking

These skills may be an optional requirement for careers in economic-administrative areas or social sciences, but in physical and mathematical areas or health sciences, they are mandatory and will allow you to develop your career successfully.

One of the most important factors is the deficiency in regulation and the lack of knowledge of local regulatory entities. But thanks to the free trade agreements with Canada and the United States (and the recent integration as a member of ICH at the end of 2021), the regulatory entity COFEPRIS (Federal Commission for the Protection against Sanitary Risks) has begun to train its staff to incorporate the Technical Requirements for the Registration of Pharmaceutical Products for human use. Mexico is the fourth country in the Americas to register with ICH, preceded by Canada, the United States, and Brazil.

From a statistical point of view, according to data obtained in the 2022 National Survey of Occupation and Employment [1], an economically active population of approximately 98 million has been registered, of which only 24 million correspond to economically active women. This is approximately 24.4% of the total economically active population.

If we review how many women are working in the pharmaceutical manufacturing sector of this active population, in 2022 there were only 55,705 women registered, the equivalent of only 0.23% of the total population of economically active women. This figure can be discouraging because here we are not only considering all women, but not just those who studied a career related to STEM.

It is here that we see the importance of sharing success stories of professional women working around biopharmaceutical medicines, whether in the private industry or academic and government research, and that their words help motivate young people in middle and high school to opt for STEM careers. In this chapter, we have conducted interviews with women who are successful and have become a benchmark in biotechnological medicines in Mexico. These women, who faced many obstacles, have gone on to manage new biopharmaceutical drugs, biosimilar drugs, and biotherapeutics.

2 Interview Dr. Helgi Jung Cook (Fig. 1)

PhD in Chemical Sciences, Faculty of Chemistry, UNAM. Member of the National System of Researchers, Level 3 (2006–2025). She has many awards and distinctions. She has more than 50 scientific publications in the last 8 years. She has directed degree and research projects for undergraduate, master's, and doctoral students. She is an active expert member of the New Molecules Committees of COFEPRIS.

As an expert in the committees of the regulations that involved topic of health, can you give us an update on those regulations in Mexico?

Dr. Jung Cook: In the case of biotechnology, the Standard is not very clear. It talks about having to carry out preclinical and clinical studies, but it leaves open

Fig. 1 Dr. Helgi Jung Cook

what the requirements would be for either of the two products, let's say for the new, the innovative, and the biocomparables. There is no clear specification, and the biggest problem has been with the biocomparables. Because there is no specification, each laboratory delivers what it has and this has made it very difficult to approve products, determine whether studies are required or not, or what type of studies they must do. Mexico specifies three batches of the comparable product versus three batches of the innovative product, but it does not specify how many relate to primary, secondary, and tertiary structures, for example. It does not indicate, at a minimum, how many tests or what type of preclinical studies should be carried out on a biocomparable. In some cases, biopharmacokinetic studies do not reflect what happens in humans. A humanized monoclonal antibody, which, if you do a study in rats, creates antibodies and pharmacokinetics that will not make much sense. So, some countries specify that they have to be in monkeys. In Mexico, within the regulations, the use of monkeys is almost impossible, and that would perhaps allow us to know receptor binding—studies that are required in any other country. That is a bit of the problem that has arisen with biocomparables. Given the lack of specifications of the studies that would have to be carried out in order for Mexico to approve them, when the dossier arrives, the committee considers that a study is missing or that the clinical study does not have a sufficient number of patients. In reality, there was no way for that laboratory to know the type of tests that are required in Mexico.

In reality, the regulations are not bad; however, we need to have clearer specifications. Europe has a very clear guide for insulins, and in Mexico, there have been more requests for biocomparable insulin products. Europeans know perfectly well what is required to be able to apply for registration. But in Mexico, as we do not have guidelines, some deliver a clinical study, some others a biopharmacokinetic study in humans, and here it seems to us that we should work a little more to guarantee quality. But I think it is good that there are biocomparables because of the cost and the possibility of social security; it has greater possibility and acquisition for the population. In order to ensure safety, quality and efficacy, the committees ultimately try to ensure the guarantee of the product. The committees are advisory; they do not give opinions (a committee determines whether the information is sufficient or not). The opinion is made by COFEPRIS, and is based on the information issued by the committee. And what the committee is most concerned about is that these molecules that are going to enter the country have quality, safety, and efficacy.

In your opinion, what benefits will the COFEPRIS agreements with the ICH bring to Mexico, how long do you think it will take before we could compare ourselves at a global level, thanks to this harmonization?

Dr. Jung Cook: The first thing that COFEPRIS is going to try to harmonize is the CTDs (Common Technical Document), the Dossiers that meet the requirements of the ICH. Many laboratories already deliver the information structured in CTD, but according to COFEPRIS, this will be for all products, not only new molecules but also generic and biosimilar medicines.

COFEPRIS considers that the electronic filing is going to take longer. With this first step it is taking, with a harmonization in CTD, it will be easier to know what the requirements are. For other guidelines that, in pharmacovigilance, are more advanced, there is another bioexemption guideline in 2025–2026. These harmonizations are already more concrete and safer.

At what age did you decide you wanted to go into science?

Dr. Jung Cook: In the fifth semester, between doing social service and starting in natural products, I decided that I would do a postgraduate degree, and I was going to dedicate myself to research, so I finished my degree. I did my master's degree in pharmacy. I got married, had children, so did not have the opportunity to travel abroad to do a doctorate. Many years later, the doctoral program opened (I think in the 90s) in pharmacy, in biopharmacy, which was what I was most interested in. I already had many publications, and in reality, the doctorate was more based on pharmacokinetics and the biopharmacy part. Now there are many more possibilities for students to enter the master's degree, or the doctorate, and to be able to concentrate on research.

When I was in my master's degree, the program called the National Biopharmaceutical Evaluation Project was opened, the PEVAB project; Dr. Espejo said to me: do you want to dedicate yourself to Biopharmacy, and I said yes. Then I had to start buying equipment that I did not know very well—they bought some dissolvers and gas chromatography equipment. Later Dr. Carlos Ramón García, who had done his doctorate in the faculty of California and started the master's program in Pharmacy, arrived. So, I migrated from the Natural Products part to the pharmacy part, and well, it has been a blessing and a challenge, as well as a pleasure to see this area change, which I like very much.

How have you seen the changes from when you started to the present, what would you say to young women who want to start their career in this line of research?

Dr. Jung Cook: Personally, this area has allowed me to develop a series of studies in our country for the pharmacokinetics part. We started with cysticercosis drugs, where there was no established dosage regimen, and it was not known if it had to be given on an empty stomach or with food, so the faculty made an agreement with the Institute of Neurology and began to do a whole range of pharmacokinetic studies. This provides the feeling that one can contribute for the benefit of the population. All the pharmacokinetic studies of that time, the work with patients, have been very pleasant, and it has left me with the feeling of being able to contribute something to the treatment of drugs to counteract cysticercosis. On the other hand, the dissolution part that has helped us to visualize how the products of the national market behave. We have been able to develop dissolution methods that did not exist or that do not exist. We are currently working with antiparasitic drugs, which is where there are fewer publications because they are not of interest in other countries. We are developing some methodologies for suspension products that we hope in the future can be included in the pharmacopoeia.

This has given us the biopharmacy part. For young women in the future, it allows them to contribute not only to supporting the quality assurance of products but also to the development of formulations that can be useful to the country and their evaluation. Many times, in the technology part, one is left in the development of the product, let's say the pharmaceutical form, but not the part of the evaluation to see if it can be useful in the clinical part that does biopharmacy and pharmacokinetics. There is a whole range of possibilities in the area of biopharmacy where graduates have an opportunity to develop. In bioequivalence, in the area of authorized third parties developing analytical methods where the laboratory has not implemented them, comes a very interesting area called Pharmacokinetic Physiological Modeling (PBPK). There are not many studies in Mexico. These are areas that must be developed to be able to be competitive with other countries that are already doing these models, to decrease human studies, and have other options for evaluating products.

There are other areas of population studies. In this area, it is about evaluating variability, why does an individual have such high concentrations and others so low? There are other covariates, such as age, gender, weight, etc., that can influence the therapeutic response. There are already several groups in Mexico that are working on population models, but it is an area that can be strengthened, that will help dosing regimens, but this is an interdisciplinary area.

3 Interview DR. MAYRA PEREZ TAPIA (Fig. 2)

Director of the Unit for Development and Research in Bioprocesses (UDIBI), Active expert member of the Committee of Biotechnological Products of the Mexican Pharmacopoeia (FEUM)

Professor-researcher "C" in the Department of Immunology of the National School of Biological Sciences (ENCB) of the National Polytechnic Institute (IPN) and professor of Immunology at the Faculty of Chemistry of the UNAM. She is a Pharmaceutical Chemist and Biologist graduated from the Faculty of Chemistry of the National Autonomous University of Mexico, with a master's and Doctorate in Science in Immunology from the ENCB of the IPN.

As executive director of UDIBI, what would be the biggest challenge you have faced in leading UDIBI, which we know is an institutional entity that depends on the IPN (Instituto Politécnico Nacional), and turning it into a company for the generation of resources?

Dr. Tapia: Science and technology were not seen as a commodity that society would need to convince decision-makers to invest, to support, and to strengthen. Areas that developed technology were always very difficult, although in the discourse we always hear that there is support in these areas, science and technol-

Fig. 2 Dr. Mayra Perez Tapia

ogy, innovation. In practice, it has been an uphill struggle to show decision-makers and investors that risks can be taken in Mexico to develop innovative antibodies, the area that is "in charge" today, or to develop pharmaceutical technologies or technologies that could become a therapy to benefit our society. The first challenge is one of knowledge.

The second challenge is normativity. The institutional regulations, nor any CONAHCYT regulations in our country, were prepared to generate technology-based companies: they were not prepared to promote entrepreneurs in science. All public universities lack, with honest exceptions, a clear path for the transfer of technology, for the management of resources of a generated good, or for the creation of companies. So, generating the regulatory platform, noting that it is beneficial to have an innovation ecosystem, is also a great challenge.

Another great challenge, since that part passed, is to develop products in a harmonious consortium. With the polytechnic and resources from CONAHCYT, you get to the regulatory part, and it turns out that the regulatory entity is not standing for innovation. We see that other countries in the world have regulation immersed in regulation; researchers approach the natural regulatory entity. The problem we face is that since COFEPRIS was created, its sanitary regulation has not been designed to innovate, so we have a regulation made to approve developments made in other countries. When researchers arrive and need to register a prototype, COFEPRIS and CONAHCYT have not contemplated it. This has been the biggest challenge that must be solved by all those who are trying to make technological innovation in the country.

The free trade agreement with the United States and Canada will help to reduce that gap—documentation and registry—so that regulations can be evened out or aligned with regulations from other parts of the world.

The moment in which we live now, after the pandemic, of a technological and geopolitical relocation in the world, is the moment in which, regardless of the free trade agreement, or international agreements, the geographical situation of Mexico is what will allow us a different interaction. For example, COFEPRIS is making alliances with several regulatory entities around the world; it belongs to ICH—not only does it have to do it out of conviction, but also because it is a commitment to the ICH—where innovation issues are involved, where they should now be included. The regulatory entity, due to our geographical position, finds it convenient to have a compliance with the United States and to have a homologation in the regulatory quality, so that our goods are in only one direction. The recognition agreements must be directional; for that to be, we must improve the quality.

As a result of the pandemic, there is a greater national awareness that something needs to be done in innovation, biotechnology, or biosimilars to stimulate national production, to have one's own supply chain, and to self-sufficiency. That knowledge of self-sufficiency that remains after the pandemic will make things move in the direction of promoting innovation in biotechnology.

As a thought leader, what strategies do you think should be promoted to break the glass ceiling in the biopharmaceutical industry to have gender equality?

Dr. Tapia: In Latin America, it is very evident; perhaps in Europe and the United States too, the gap is less evident in gender equality.

Looking at it from a geographical point of view, from the point of view of one hemisphere and another, and of how women are positioned for this type of management position: if you want to change something as important as the management of companies, then you are talking about women who must have technical skills. Mexico is migrating. The next president of this country will be a woman, regardless of the color of the party, it will be a woman, which will make us advance three steps, as in the boxes of board games. It is already a powerful message for girls and women that if we can reach the maximum point of position of a woman in the country of being president of the Republic, but the everyday speech, which one listens to from home, should repeat this message—it must be everyday speech.

Women, by nature, have certain physiological breaks for those of us who take the time to be mothers, because you also live for this wonderful activity. It is valid to stop and complement your life with motherhood. It is fundamental, with the maternal presence, in the first years of life, there is a dependence of the baby on the mother. No matter how many partners or supportive husbands who help and get up to give a bottle, maternal dependence forces women to have spaces in their working life that are different from those of men. Fortunately, now, responsible fatherhood has increased its presence and is now going to help standardize the paths.

In summary, what needs to be done so that there are more women in management positions is a change in the global policy of positioning women. A global discourse must be made of greater belonging and the importance of the female entity.

Second, integrating such policies within organizations and public policies, within organizations, not only by number but also by action, is very important.

Third, to give greater possibilities in countries like ours where the gap is larger, it is not necessary for women to have priority. If we have not been for dozens of years with respect to men, to reduce the gap, we have to give greater facilities to women. Perhaps in 10 years we will have parity.

Fourth, and without a doubt, to change at the political level the working conditions of women who decide to prepare and continue their training. In Mexico, for example, doctoral students do not have access to public child care. If a doctoral student combines her research or doctoral activity with a pregnancy, there are no nurseries that allow her to have access to child care. We have to change the public policy of how we support the preparation of women in this critical reproductive stage for those who decide to reproduce, because now pregnancy is not by obligation, it is a choice.

Is being a woman in biotechnology complicated? What are the challenges you face daily?

Dr. Tapia: The challenges have depended on the stage. When my daughters were young, she has three teenage daughters, the main challenge was organization—how to balance the time to dedicate yourself as a person, as a mother, as a family, a professional, and give yourself time to take care of yourself. As a person, as a woman, the main challenge is to maintain a balance. Sometimes the same situation on a day-to-day basis leans toward a maternal side. I have to be present because it is the season of exams, graduations, etc. I have to see the future activities of my daughters: school, registrations, but there are also other moments when I am completely working because it is the closing of projects. You have to set budgets; you have to work day and night. The main challenge is maintaining a balance, where you don't feel guilty for attending to another. That is challenge one.

Challenge number two is to work against the inertia of the system. It is difficult to work against the inertia of the system, that is, you have a dynamic in the pharmaceutical industry, a dynamic of innovation, a dynamic of it must be for today because tomorrow someone else is already thinking the same as you… let's hurry. Working within the inertia of a Mexican bureaucratic system, which is wonderful and has given me all the possibilities but is still bureaucratic, and trying to transmit that sense of positive urgency has been difficult. It's been a challenge on a day-to-day basis.

Third, to do science in Mexico, to do science and Innovation in Mexico, (few people know that we pay for scientific development about 30% more than what it costs in the United States or what it costs in Europe), there is little love, and we waste it on jealousy. Investment in science is already low, and it is also more expensive than in the United States, which is our neighbor and is our reference source. As a very simple example, in the United States, you order a protein from a supplier and at night it is already in the laboratory. In Mexico, the same protein can arrive after 2 months. The value of that time is incalculable when you compare prices. And there is an intermediary, so they are paying practically double the value of the protein. We are paying double, and the arrival time is late. There is no free situation of the loan system that they traditionally have in the United States; in Mexico, it is more difficult to execute.

Fourth, the generational difference. Generations have changed so quickly. I am a teacher at the university and the polytechnic, and I also work with a lot of young people, so you have to adapt to the thinking of these new generations, and it may be that this new thinking does not coincide with your way of thinking. So, it has been a great challenge. You have to listen to young people; they have new goals, new ways of communicating, other forms of motivation, even other ways of learning, and other ways of following leadership. Today, a good boss is not only the boss who knows—the boss who knows the one who is on Tik Tok, the one who's present, the one who does things—that's what kids are looking for nowadays. When you understand the position of leader, it's a lot of responsibility because you have to devote time according to what those people need from you as a leader, give you the opportunity to prepare, to take courses, and to adapt to this new situation. I am one of the oldest in my organization, and I am 50 years old. The average age in my organization is under 30 years old. I have a lot of young people graduating from universities, and I am forced to do so. Mayra is passionate about her work; it is really very interesting. In addition to the simple fact of doing something that you like, such as innovation and biotechnology, doing complex projects, there are many variables that you have to take care of. There are many actors that must be taken care of, and many situations that depend on bureaucracy that do not depend on you. You have to go to the boss so that he can facilitate or make an exception for 25 procedures, or an express review so that you comply with your deadlines—it is a motivation to be day-to-day.

From what age did you envision a career in biotechnology?

Dr. Tapia: My story is very applicable because I grew up in a laboratory. The memories that come to my mind when I enter a laboratory—when I smell alcohol, solvents, when I hear a centrifuge—are memories of my childhood of a profoundly happy childhood. That undoubtedly marks what you are going to address in the future, without a doubt. My mother is a chemist in a laboratory of clinical analysis. At the time, it was initially set up in a small town in Veracruz near Los Tuxtlas. Since I was very young, since I was conscious of going down to the laboratory, I knew that I had to dedicate myself to some of the things related to bio.

Another thing that marked my future, although I went to study in the port of Veracruz and then in Mexico City, is that I always return to my town, to my small city. It is a sugarcane city, and a city nestled in the Tuxtlas, which are full of gifts of nature that are used in herbalism. I remember that I lived on the second floor, and I saw a merolico (a vendor) who stopped every Wednesday morning and brought a lot of herbs, an herbalist. He put them on the floor, and he brought a lot of parasites, samples of earthworms (*Taenia* sp.), his mixtures of specimens, along with all his plants. I remember the pleasure it gave me to visit with that man. I was very young, about 10 years old, to sit down and ask him about the plants, to smell them. And to tell me about the remedies—that there was a dewormer, that there was one for headaches, that there was another to dissolve kidney stones, and there was another for migraines; and each of those herbs smelled different, had a different vision and use, and smelled different. I remember telling my mom, that's not magical, there must be something that works for that, and I learned.

When I had to leave my home to go to the port of Veracruz, I was very clear that I wanted to do something about biology, something about development, something about the area of science. In high school, I was tempted by physics—I have an aunt who is a chemist and nuclear chemist who worked in Laguna Verde at the time—and liked that very much. I had one of those teachers who touch the soul, who was very good, so I was between physics and chemistry, but I opted for chemistry because I was a QFB (Químico Fármaco-Biólogo). I applied to the University of Veracruz, I applied to the UNAM, and I studied at the UNAM. It was at a very early stage when I decided my vocation. And today, when I see kids who are 18–20 years old and have not defined their vocation because there are so many things to do, I feel deeply fortunate to have made that choice at such a young age. After that, I am very attracted to the issue of technology-based management, innovation, and biotechnology companies. But the technical base is the one that helps without a doubt, and I couldn't do the other if I didn't study the former, but it was very early that I was first influenced by family. I don't know what would have happened if I hadn't grown up in that environment.

As a society, how could we support encouraging women to pursue a career in science, technology, engineering, or mathematics?

Dr. Tapia: There are many ways. The first is public policy; we must have an impact, so that there is greater support for girls who want to study in this way. That is, specific support—you want to study mathematics, you have a scholarship. It does not end just because you are a woman. But there must be a strong, aggressive public policy to support women who decide to dedicate themselves to it.

Dissemination is very important. We must reach a greater community of girls and adolescents, and that is a great challenge because, unfortunately, the opinion leaders are the ones who make bags, sing, and walk half-naked in the street, which I do not say is right or wrong, as opposed to scientists. But we need more dedicated influencers to talk about this, influencers who are really convinced that what they say about science and technology is going to be heard by society, by that critical mass of girls, who must see science, mathematics, and engineering as a possibility and as something where you can have skills. One thing is what you like and another is what you are good at. Try to look for things that match. If your brain and your characteristics are telling you that you are good at math, fashion, the stock market, being a fashion designer, and being an influencer, please try to match passion with skill. Because if we don't, we'll put our foot in it; it's a logical thing to do.

Public policy support, greater dissemination, and identifying spokespersons who can reach that community of children or adolescents who are glued to their cell phones. Young people have grown up with them, and we must have good communication.

Spokespersons should be identified who can reach the girls now, in a way of communication that is attached to the mobile phone. This is the way of communication. It is a necessary evil, but it is the way we communicate today, and we use it as a work instrument. Of course, it has also taken us away from other areas of life. It is necessary to communicate.

Another thing that motivates me, and that I have realized that it does have an effect, is when you go directly to the areas, where there are girls. For example, I have gone several times to schools in Sinaloa, Mazatlán, Culiacan, Tijuana, Veracruz, and Guerrero, where I go and give lectures to high school girls. The simple fact that you make yourself visible, that you become someone available, someone palpable, someone who goes and tells them, this is possible. Also, coming from the province and understanding that at that age I also had my desires (it was different because there was not so much multimedia), but I think it is very important. It has a greater impact than one might believe. I have received students who come and later send a message: Dr. Tapia, you went to a conference a few years ago, I listened to it, and now I am a chemist, or Dr., you visited, and now I am studying science, and I am going to La Joya! You don't realize it, but the impact you leave on those girls is very relevant; it really changes lives. Middle and high school are the ideal time to touch lives and promote the sciences.

What would you say to girls, and young women who have a vocation in health sciences to encourage them to pursue a career in biotechnology?

Dr. Tapia: Above all, for young people (I have teenagers, and I see their friends), I have tried to be a very present mother. One of the things that I realize, and not only in a social situation, is the feeling of immediacy—this hurts young people the most today, this immediacy. So, the first thing I see when I give classes or a conference, I try to motivate kids. The first thing I talk to them about is immediacy, the advantages and disadvantages of having everything immediately and the awareness of what to expect from the culture of effort, the culture of patience. It's the first thing. I try to give examples—immediacy gives you momentary pleasure, it gives you immediate satisfaction, it gives you belonging to something you're not—but the long-term satisfaction of being able to do something for others, to be able to develop the ability to take care of yourself, for your grandmother, your mother, for yourself; that you can leave a legacy to humanity, gives you a satisfaction that cannot be compared. It is an intangible that you must teach—that value in difference.

There is a prejudice. Some guys told me, you are going to be a scientist, you are going to die of hunger, sorry gentlemen, but thank God I have never starved to death! All the people who work with passion in what they do, with good ideas and preparation, 99% of people are successful because they look for a way to achieve their goals. What I tell them is that the culture of effort is one that has the greatest rewards more than the culture of immediacy, and that it is possible. And while they do not realize it at that moment, I can be an example. I left my home at the age of 14 to arrive in a different city; from a small bicycle city, and then to a Port, and then migration to the big City (without knowing anyone). I had never ridden the subway. The first and second time I got on the subway alone was the first day of university. The rewards began to arrive soon, and the reward has been growing and growing. The impact of my work is greater than a number of people, so there is a lot of satisfaction.

An example I give to girls and women, including colleagues and friends, is to learn—to learn to do something for others in community service and to see the joy that it gives you to do something for the benefit of another. When you dedicate yourself to biotechnology, you are doing that, but you're doing it on a macro level. Those who developed the vaccines that saved us from the pandemic saved humanity and left a legacy that is going to be indelible in all the history books. That's an incalculable satisfaction. Biotechnology gives you that—biotechnology, veterinary, agricultural, or health. The simple fact of thinking that you're developing something that can be developed and potentially used to improve life is growth, and it's legacy.

One of the things that I've tried to communicate all my life is that your first development is not necessarily going to be successful, but along the path of development, there is a lot of learning. For example, many people criticize that the national vaccine that is being developed is already late, but the learning that is being generated and the path that is being developed are things that may not be realized today but bring a lot of value for the future. Learning to communicate that there is a lot of value in medium- and long-term efforts in the implementation of science, in development, and in being able to do things for others is important. Developing a new biotechnology or a new process and being able to improve or optimize—improving water, creating a new drug, making a vaccine—thinking of how that's for the social good, is important.

Looking forward 10 years, with a different ecosystem, it is critical to encourage more women to work (and not feel pressured to leave due to family responsibilities), replicate positive examples, and let women know they share a cause of support for more women. You will always see inequalities, and you will always see a woman to support.

Reference

1. 2022 National Survey of Occupation and Employment, managed by INEGI (National Institute of Statistics, Geography and Informatics). https://en.www.inegi.org.mx/programas/enoe/15ymas/

Index

S. Taylor, M. Campbell (eds.), *Women in Biopharma*, Women in Engineering and Science, https://doi.org/10.1007/978-3-031-80348-2

The manufacturer's authorised representative in the EU is Springer Nature Customer Service Centre GmbH, Europaplatz 3, 69115 Heidelberg, Germany. If you have any concerns regarding our products, please contact ProductSafety@springernature.com

Printed and bound by CPI Group (UK) Ltd, Croydon, CR0 4YY
01/10/2026
02225977-0015